The Canadian Institute
of Resources Law



# THE OFFSHORE PETROLEUM REGIMES OF CANADA AND AUSTRALIA

Constance D. Hunt
Executive Director
Canadian Institute of Resources Law
and
Professor of Law
Faculty of Law
The University of Calgary
Calgary, Alberta

June 1989

Faculty of Law. The University of Calgary, Calgary, Alberta. T2N 1N4 Tel. (403) 220-3200

**Canadian Cataloguing in Publication Data**

Hunt, Constance D. (Constance Darlene), 1950-
The offshore petroleum regimes of Canada and Australia

ISBN 0-919269-29-X

1. Oil well drilling, Submarine - Law and legislation - Canada. 2. Oil well drilling, Submarine - Law and legislation - Australia. 3. Offshore oil industry - Canada. 4. Offshore oil industry - Australia. I. Canadian Institute of Resources Law. II. Title.
KE1815.H85 1989 343.71'0772 C89-091277-7

## THE CANADIAN INSTITUTE OF RESOURCES LAW

The Canadian Institute of Resources Law was incorporated in 1979 with a mandate to examine the legal aspects of both renewable and nonrenewable resources. Its work falls into three interrelated areas: research, education, and publication.

The Institute has engaged in a wide variety of research projects, including studies on oil and gas, mining, forestry, water, electricity, the environment, aboriginal rights, surface rights, and the trade of Canada's natural resources.

The education function of the Institute is pursued by sponsoring conferences and short courses on particular topical aspects of resources law, and through teaching in the Faculty of Law at The University of Calgary.

The major publication of the Institute is its ongoing looseleaf service, the *Canada Energy Law Service*, published in association with Richard De Boo Publishers. The results of other Institute research are published as books and discussion papers. Manuscripts submitted by outside authors are considered. The Institute publishes a quarterly newsletter, *Resources*.

The Institute is supported by the Alberta Law Foundation, the Government of Alberta, the Government of Canada, The Foundation for Legal Research, and the private sector. The members of the Board of Directors are appointed by the Faculty of Law at The University of Calgary, the President of The University of Calgary, the President of the Law Society of Alberta, the President of the Canadian Petroleum Law Foundation, and the Dean of Law at The University of Alberta. Additional members of the Board are elected by the appointed Directors.

All enquiries should be addressed to:

The Executive Director
Canadian Institute of Resources Law
Room 430, BioSciences Building
Faculty of Law
The University of Calgary
Calgary, Alberta, Canada
T2N 1N4

Telephone: (403) 220-3200
Telex: 03-821545
Facsimile: (403) 282-8325

# LE CANADIAN INSTITUTE OF RESOURCES LAW

Le Canadian Institute of Resources Law a été constitué en 1979 et a reçu pour mission d'étudier les aspects juridiques des ressources renouvelables et non renouvelables. Son travail porte sur trois domaines étroitement reliés entre eux, soit la recherche, l'enseignement et les publications.

L'institut a entrepris une vaste gamme de projets de recherche juridique, notamment des études portant sur le pétrole et le gaz, l'exploitation des mines, l'exploitation forestière, les eaux, l'électricité, l'environnement, les droits des autochtrones, les droits de surface et le commerce des ressources naturelles du Canada.

L'institut remplit ses fonctions éducatives en commanditant des conférences et des cours de courte durée sur des sujets d'actualité particuliers en droit des ressources et par le truchement de l'enseignement à la Faculté de droit de l'université de Calgary.

La plus importante publication de l'institut est son service de publication continue à feuilles mobiles intitulé le *Canada Energy Law Service*, publié conjointement avec Richard De Boo Publishers. L'institut publie également les résultats d'autres recherches sous forme de livres et de documents de travail. Les manuscrits soumis par des auteurs de l'extérieur sont également considérés. L'institut publie un bulletin trimestriel intitulé *Resources.*

L'institut reçoit des subventions de la Alberta Law Foundation, du gouvernement de l'Alberta, du gouvernement du Canada, de la Fondation pour la recherche juridique et du secteur privé. Les membres du conseil d'administration sont nommés par la Faculté de droit de l'université de Calgary, le recteur de l'université de Calgary, le président de la Law Society of Alberta, le président de la Canadian Petroleum Law Foundation et le doyen de la Faculté de droit de l'université d'Alberta. D'autres membres sont élus par les membres du conseil nommés.

Toute demande de renseignement doit être adressée à la

Directrice générale
Canadian Institute of Resources Law
Room 430, BioSciences Building
Faculty of Law
The University of Calgary
Calgary, Alberta, Canada
T2N 1N4

Téléphone: (403) 220-3200
Télex: 03-821545
Facsimilé: (403) 282-8325

## TABLE OF CONTENTS

# FOREWORD

This study of Canadian and Australian legal and administrative arrangements for the management of offshore petroleum laws was carried out from September 1987 to September 1988, during my sabbatical leave from The University of Calgary and from my position as Executive Director of the Canadian Institute of Resources Law. The project received financial support from the Universities of Calgary and Melbourne and the Social Sciences and Humanities Council of Canada, all of which is gratefully acknowledged.

During the fall of 1987, research was carried out on the Canadian part of the study in Calgary, Ottawa, Halifax and St. John's, including interviews with personnel in industry and government and with members and staff of the Canada -Newfoundland Offshore Petroleum Board. Research on the Australian material was conducted during the first four months of 1988, during a period of residence at the University of Melbourne. The research in Australia also included interviews with government and industry officials in Melbourne, Canberra, Sydney, Adelaide, and Perth.

The interviews provided some of the most important information for the study, but, in conformity with the policies and requirements of The University of Calgary Ethics Committee, individuals have been quoted directly only when their specific consent was sought and obtained. The information procured through interviews has mostly been generalized, without ascribing particular comments or views to one individual.

I received assistance from many people, including Professor Alastair Lucas, The University of Calgary; Professor Michael Crommelin, University of Melbourne; Professor Richard Cullen, Monash University; Maurice Taschereau, Dan Whelan, Brian Moore, Tamara Parschin-Rybkin, Wayne Lonsdale, Ken Sato, Graham Campbell, Glen Yungblut, and Bill Smith, COGLA; George Anderson, Department of Energy, Mines and Resources,

Government of Canada; John McWilliams, formerly of Mobil Oil Canada, Ltd.; Don Bews, John Andrews, Wes Abel, and Stu Aitken, Mobil Oil Canada, Ltd.; Jim Pasieka, Pat Purcell, and Ferg Willmer, Husky Oil Operations Ltd.; Ted Baugh, John Fitzgerald, Diego Henao, and Angus Taylor, Canada-Newfoundland Offshore Petroleum Board; Gordon Gosse and Martin Shepherd, Department of Energy; Ron Stevenson, Justice Department and Ronald Penney, Department of Public Works and Services, Government of Newfoundland and Labrador; Peter Outhit, Nova Scotia Resources Limited; Laurie Taylor, Walter Muscoby and G.F. Smith, Texaco Canada Resources; Andrew Parker and Sara Jane Snook, Department of Mines and Energy, and Doug Keefe, Department of Justice, Government of Nova Scotia; Rowland Harrison, Stikeman Elliott; P.D. Bretzloff, K. MacFarlane, J. Nesbitt, D. Schroeder, and S. Jim Arsenych, Petro-Canada; Ken Gardner, Department of Industry, Technology and Resources, Government of Victoria; Bill Mason and Ian Fraser, Department of Mines, and Robin Griffiths, Crown Law Department, Government of Western Australia; Bryan Griffith, John Grace, and Michael Woollands, BHP Petroleum; Andrew Thompson, Michael Wright, Michael Knowles and Richard Rowick, Mallesons Stephen Jacques; Peter Reid, Corrs Pavey Whiting and Byrne; Frank Hooke, Baker and McKenzie; Richard Kerlin, Marathon Petroleum Australia, Ltd.; Clem Allworth, Woodside Offshore Petroleum Pty. Ltd.; Robin Maxam, Bond Corporation; Humphrey Williams, West Australia Petroleum; Rory Argyle, Parker and Parker; Warren Bennett, Australian Petroleum Exploration Association Limited; Robert Laws and Graham Herrmann, Department of Mines and Energy, Government of South Australia; Dennis Ives, John Tilley, Mike Hitchens, J.C. Starkey, and John Carlson, Department of Primary Industries and Energy, and Henry Burmester, Attorney-General's Department, Australian Government.

I also wish to thank Milva Bello, Nancy Money, Sue Parsons and Cathy Price for first-rate support in typing this manuscript. Research assistance has been provided by Donna Maxwell and by

Cecilia Lowe, who also helped design the tables. Evangeline Case's contribution as editor is gratefully acknowledged.

This study forms part of the Canada Lands Project, which has been sponsored by Chevron Canada Resources Limited, Esso Resources Canada Limited, Gulf Canada Corporation, Home Oil Company Limited, Husky Oil Operations Ltd., Mobil Oil Canada Ltd., Petro-Canada Inc., Shell Canada Limited, and Suncor Inc. Resources Group. Support for the project has also been provided by the Department of Indian Affairs and Northern Development.

Constance D. Hunt
Executive Director
Canadian Institute of Resources Law

# AVANT-PROPOS

Cette étude des ententes juridiques et administratives canadiennes et australiennes sur la gestion des lois afférentes au pétrole au large des côte a été réalisée de septembre 1987 à septembre 1988, au cours de mon congé sabbatique de l'université de Calgary et de mon poste de directrice exécutive du Canadian Institute of Resources Law. Ce projet a été financé par les universités de Calgary et de Melbourne ainsi que par le Conseil de recherches en sciences humaines du Canada, et je les remercie de leur apport.

La recherche sur la partie canadienne de cette étude a été faite au cours de l'automne 1987 à Calgary, Ottawa, Halifax, et St. John's, y compris les entrevues avec des membres de l'industrie et des gouvernments ainsi qu'avec des membres et du personnel de l'Office Canada-Terre-Neuve des hydrocarbures extracôtiers. La recherche sur la documentation australienne a été réalisée pendant les quatre premiers mois de 1988, au cours d'un séjour à l'Université de Melbourne. Cette recherche comprenait également des entrevues avec des représentants du gouvernement et de l'industrie à Melbourne, Canberra, Sydney, Adelaide et Perth.

Quelques-uns des renseignements les plus importants pour l'étude ont été recueillis lors des entrevues, mais conformément aux politiques et exigences du University of Calgary Ethics Committee (Comité d'éthique de l'Université de Calgary), les individus n'ont été cités directement que lorsque leur consentement express avait été obtenu. Les renseignements obtenus grâce aux entrevues ont été dans l'ensemble généralisés, sans attribuer de commentaires ou de points de vue particuliers à un individu spécifique.

De nombreuses personnes m'ont offert leur appui, notament le professeur Alastair Lucas de l'Université de Calgary; le professeur Michael Crommelin de l'Université de Melbourne; le professeur Richard Cullen de l'Université Monash; Maurice Taschereau, Dan Whelan, Brian Moore, Tamara Parschin-Rybkin, Wayne Lonsdale,

Ken Sato, Graham Campbell, Glen Yungblut et Bill Smith de l'Aministration du pétrole et du gaz des terres du Canada; George Anderson du ministère de l'Énergie, des Mines et des Ressources du gouvernement du Canada, Ltd.; John McWilliams, auparavant avec Mobil Oil Canada, Ltd.; Don Bews, John Andrews, Wes Abel et Stu Aitken de Mobil Oil Canada, Ltd.; Jim Pasieka, Pat Purcell et Ferg Willmer de Husky Oil Operations Ltd.; Ted Baugh, John Fitzgerald, Diego Henao et Angus Taylor de l'Office Canada-Terre-Neuve des hydrocarbures extracôtiers; Gordon Gosse et Martin Shepherd du ministère de l'Énergie; Ron Stevenson du Justice Department (ministère de la Justice) et Ronald Penney du Department of Public Works and Services (ministère des Travaux et Services publics) du gouvernement de Terre-Neuve et du Labrador; Peter Outhit de Nova Scotia Resources Ltd.; Laurie Taylor, Walter Muscoby et G.F. Smith de Texaco Canada Resources; Andrew Parker et Sara Jane Snook du Department of Mines and Energy (ministère des Mines et de l'Energie) et Doug Keefe du Department of Justice (ministère de la Justice) du gouvernement de la Nouvelle-Écosse; Rowland Harrison de Stikeman Elliott; P.D. Bretzloff, K. MacFarlane, J. Nesbitt, D. Schroeder et S. Jim Arsenych de Petro-Canada; Ken Gardner du Department of Industry, Technology and Resources, gouvernement de Victoria; Bill Mason et Ian Fraser du Department of Mines et Robin Griffiths du Crown Law Department, gouvernement de l'Australie Occidentale; Bryan Griffith, John Grace et Michael Woollands de BHP Petroleum; Andrew Thompson, Michael Wright, Michael Knowles et Richard Rowick de Mallesons Stephen Jacques; Peter Reid de Corrs Pavey Whiting et Byrne; Frank Hooke de Baker and McKenzie; Richard Kerlin de Marthon Petroleum Australia, Ltd.; Clem Allworth de Woodside Offshore Petroleum Pty. Ltd.; Robyn Maxam de Bond Corporation; Humphrey Williams de West Australia Petroleum; Rory Argyle de Parker and Parker; Warren Bennett de la Australian Petroleum Exploration Association Limited; Robert Laws et Graham Herrmann du Department of Mines and Energy, gouvernement de l'Australie Méridionale; Dennis Ives, John Tilley, Mike Hitchens, J.C. Starkey et John Carlson du Department of Primary Industries and Energy et Henry Burmester du Attorney-General's Department, gouvernement de l'Australie.

Je remercie également Milva Bello, Nancy Money, Sue Parsons et Cathy Price qui ont fait un travail de premier ordre en tapant ce manuscrit. L'aide à la recherche a été fournie par Donna Maxwell et Cecilia Lowe, qui ont également aidé à réaliser les tableaux. Je remercie enfin Evangeline Case pour son travail d'édition.

Cette étude fait partie du projet sur le droit du pétrole et du gaz sur les terres canadiennes, qui a été financé Chevron Canada Resources Limited, Esso Ressources Canada Limitée, Corporation Gulf Canada, Home Oil Company Limited, Husky Oil Operations Ltd., Mobil Oil Canada Ltd., Petro-Canada Inc., Shell Canada Limitée et Suncor Inc. Resources Group. Le ministère des Affaires indiennes et du Nord canadien a également fourni un appui au projet.

Constance D. Hunt
Directrice exécutive
Canadian Institute of Resources Law

# ABSTRACT

As federal states, both Canada and Australia have struggled to cope with the fact that two levels of government wish to play a role in the management of offshore petroleum resources. While litigation has had an influence, in each country solutions have been achieved through intergovernmental agreements implemented by federal-provincial/state legislation. This study reviews the approaches taken in the two countries with a view to their similarities and differences, their impact on industry, and what lessons may be learned from each.

Despite the emergence of offshore jurisdictional disputes in Canada in the 1960s, it was not until 1982 that Canada and Nova Scotia agreed upon joint management arrangements. Although the Newfoundland-Canada dispute was later decided by the courts, those two governments also agreed to a joint management regime in 1985 (the *Atlantic Accord*). This regime provided a model for a new arrangement between Canada and Nova Scotia in 1986 (the *Nova Scotia Accord*). Both Accords contemplate the establishment of federal-provincial boards to oversee offshore petroleum management and recognize provincial legislative authority in relation to certain fiscal and social matters.

Offshore jurisdictional disputes also occurred in Australia in the 1960s, leading to the *1967 Agreement* between the Commonwealth and all state governments. This arrangement, implemented by "mirror" legislation passed by both levels of government, recognized considerable state authority and provided for revenue-sharing between the governments. In 1973, the Commonwealth passed legislation asserting its authority beyond the low water-mark, legislation that was unsuccessfully challenged by the states in the High Court. As a result of a new intergovernmental arrangement in 1979, the states were granted rights to and authority over a three-mile offshore area. Commonwealth petroleum legislation applies to the "adjacent areas" beyond three miles; state authorities handle day-to-day administration of the adjacent areas, but important decisions are made by a "Joint Authority" made up of the federal and state ministers. Offshore petroleum revenues continue to be shared.

Several themes common to both countries are addressed in this study. These include the legal and constitutional status of the political Accords and what effect such status has had on the ongoing relationships between the levels of governments; the procedures by which the enabling laws and their regulations are amended; the extent to which it has been possible to achieve consistency in the laws, regulations, and policies that pertain to the various offshore areas in each country; the workability of the respective decision-making mechanisms; the clarity of legislative authority under each regime; fiscal relationships; and the legal and practical significance of boundaries in the offshore areas.

The conclusions focus upon the ways in which future Canadian offshore regimes might be improved as a result of the experiences in both countries.

## RÉSUMÉ

En tant qu'états fédéraux, le Canada et l'Australie ont eu beaucoup de difficulté à accepter le fait que les deux paliers de gouvernement veulent jouer un rôle dans la gestion des ressources pétrolières situées au large des côtes. Bien que les tribunaux aient été appelés à intervenir, chaque pays est arrivé à une solution en recourrant à des ententes intergouvernementales appliquées par les lois fédérales et provinciales/des états. Cette étude examine les approches adoptées par les deux pays en analysant leurs ressemblances et leurs différences, leur impact sur l'industrie, et les leçons qu l'on peut tirer de chacune d'elles.

Bien que des conflits relatifs à la juridiction au large des côtes aient surgi au Canada dès les années 1960, ce n'est qu'en 1982 que le Canada et la Nouvelle-Écosse se mirent d'accord sur des ententes de gestion conjointe. Bien que le conflit entre le Canada et Terre-Neuve fut plus tard résolu judiciairement, ces deux gouvernements acceptèrent également un régime de gestion conjointe en 1985 (l'Accord Atlantique). Ce régime servit de modèle à une nouvelle entente entre le Canada et la Nouvelle-Écosse en 1986 (l'Accord de la Nouvelle-Écosse). Ces deux Accords prévoient la création de commissions fédérales-provinciales chargées de surveiller la gestion du pétrole au large des côtes et reconnaissent l'autorité législative provinciale au regard de certaines matières fiscales et sociales.

Des conflits relatifs à juridiction au large des côtes surgirent aussi en Australie dans les années 1960 et aboutirent à une entente entre le Commonwealth et tous les gouvernements des États. Cette entente, appliquée par des lois identiques adoptées par les deux paliers de gouvernement, reconnaissait aux États une autorité considérable et prévoyait le partage des recettes entre les gouvernements. En 1973, le Commonwealth adopta une loi affirmant son autorité au-delà de la ligne des basses eaux. Cette loi fut attaquée sans succès par le États devant la Haute Cour. A la suite d'une nouvelle entente intergouvernementale en 1979, les

États obtinrent des droits et se virent reconnaître compétence sur une zone de trois milles au large des côtes. Les lois du Commonwealth régissant le pétrole s'appliquent aux "zones adjacentes" situées au-delà des trois milles; les autorités des États sont chargées de l'administration journalière des zones adjacentes, mais les décisions importantes sont prises par une "Joint Authority" constituée de ministres fédéraux et de ministres des États. Les recettes provenant du pétrole situé au large des côtes continuent à être partagés.

Cette étude examine plusieurs thèmes communs aux deux pays, notamment le statut juridique et constitutionnel des accords politiques et l'effet que ce statut a eu sur les relations suivies entre les deux paliers de gouvernement; les procédures selon lesquelles les lois d'habilitation et leurs règlements sont modifiées; le degré d'uniformité des lois, règlements et politiques régissant les diverses zones extracôtières de chaque pays; la réussite des mécanismes de prise de décisions respectifs; la clarté de la compétence législative dans chaque régime; les rapports fiscaux, et l'importance juridique et pratique de frontières dans les zones extracôtières.

En conclusion, l'étude examine la façon dont les régimes canadiens applicables au large des côtes pourraient être améliorés à l'avenir en utilisant les expériences des deux pays.

# TABLE OF ABBREVIATIONS

## AUSTRALIA

| | |
|---|---|
| AAT | Administrative Appeals Tribunal |
| AMEC | Australian Minerals and Energy Council (and its Standing Committee on Offshore Petroleum Legislation) |
| BHP | Broken Hill Proprietary Company Ltd. |
| BMR | Bureau of Mineral Resources |
| DA | Designated Authority (state) |
| DRE | Department of Resources and Energy |
| JA | Joint Authority (Commonwealth) |
| OCS | *Offshore Constitutional Settlement* |
| PSLA | *Petroleum (Submerged Lands) Act 1967* (Cth) |
| RRR | Resources Tent Royalty |
| RRT | Resource Rent Tax |

## CANADA

| | |
|---|---|
| CNOPB | Canada-Newfoundland Offshore Petroleum Board |
| CNSOPB | Canada - Nova Scotia Offshore Petroleum Board |
| COGLA | Canada Oil and Gas Lands Administration |
| CPRA | *Canada Petroleum Resources Act* |
| FD | Fundamental Decision |
| MODUS | Mobile Offshore Drilling Units |

| | |
|---|---|
| MOU | Memoranda of Understanding |
| OGPCA | *Oil and Gas Production and Conservation Act* |

# TABLE OF AGREEMENTS

## AUSTRALIA

"Agreed Arrangements",
*in* Australia, Dept. of the Attorney-General, *Offshore Constitutional Settlement: A Milestone in Co-operative Federalism* (Canberra: Australian Government Publishing Service, 1980) reproduced in part *in* Michael Crommelin, "Offshore Mining and Petroleum Constitutional Issues" (1981) 3 A.M.P.L.J. 191 at Appendix A

Agreement relating to the Exploration for, and the Exploitation of, the Petroleum Resources, and certain other Resources, of the Continental Shelf of Australia and of certain Territories of the Commonwealth and of certain other Submerged Lands,
reprinted *in* Australia, Senate Select Committee, *Report from the Senate Select Committee on Offshore Petroleum Resources* (Canberra: Commonwealth Government Printing Office, 1971) at Appendix A [hereinafter *1967 Agreement*]

Barrow Island Agreement

1967 Agreement
see
Agreement relating to the Exploration for, and the Exploitation of, the Petroleum Resources, and certain other Resources, of the Continental Shelf of Australia and of certain Territories of the Commonwealth and of certain other Submerged Lands

1979 OCS
see
1979 Offshore Constitutional Settlement

1979 Offshore Constitutional Settlement
*in* the information kit, entitled "Offshore Australia", which includes maps and two booklets: Australia, Dept. of the

Attorney General, *Offshore Constitutional Settlement: A Milestone in Co-operative Federalism* (Canberra: Australian Government Publishing Service, 1980) (reproduced in part in Michael Crommelin, "Offshore Mining and Petroleum Constitutional Issues" (1981) 3 A.M.P.L.J. 191) and Australia, Dept. of the Attorney General, *Offshore Constitutional Settlement: Selected statements and documents.*

**CANADA**

Atlantic Accord

Meech Lake Constitutional Accord

Memorandum of Understanding in Respect of the Administration and Management of Mineral Resources Offshore of the Maritime Provinces [hereinafter 1977 Understanding]

Northwest Territories Northern Accord Agreement in Principle

Nova Scotia Accord

Yukon Northern Accord Agreement in Principle

1964 Agreement
(between the Premiers of Nova Scotia, New Brunswick, Newfoundland, Prince Edward Island, and Quebec concerning offshore boundaries)

1977 Understanding
see
Memorandum of Understanding in Respect of the Administration and Management of Mineral Resources Offshore of the Maritime Provinces

1982 Canada - Nova Scotia Agreement
[hereinafter 1982 Nova Scotia Agreement]

1982 Nova Scotia Agreement
see
1982 Canada - Nova Scotia Agreement

## TABLE OF CASES

### AUSTRALIA

*BHP Petroleum Pty. Ltd.* v. *Balfour*,
[1987] 61 A.L.J.R. 345 (Aust. H.C.)

*Fordham and the State of Victoria* v. *Evans*
(13 November 1987), No. VG428 of 1986 (Aust. F.C.)

*Hematite Petroleum Pty. Ltd.* v. *State of Victoria*
(1983), 151 C.L.R. 599 (Aust. H.C.)

*New South Wales* v. *The Commonwealth*
(1975), 135 C.L.R. 337 (Aust. H.C.)
[hereinafter *Seas and Submerged Lands Case*]

*Seas and Submerged Lands Case*
see
*New South Wales* v. *The Commonwealth*

### CANADA

*A.G. Canada* v. *A.G. British Columbia*,
[1984] 4 W.W.R. 289 (S.C.C.) (*sub nom. Reference re the Ownership of the Bed of the Strait of Georgia and Related Areas*)

*B.C. Offshore Reference*
see
*Reference re Offshore Mineral Rights of British Columbia*

*Hibernia Reference*
see
*Reference re the Seabed and Subsoil of the Continental Shelf Offshore Newfoundland*

*Newfoundland Reference*
see
*Reference re Mineral and Other Natural Resources of the Continental Shelf*

*Reference re Mineral and Other Natural Resources of the Continental Shelf*
(1983), 145 D.L.R. (3d) 9 (Nfld. C.A.)
[hereinafter *Newfoundland Reference*]

*Reference re Offshore Mineral Rights of British Columbia*,
[1967] S.C.R. 792, 62 W.W.R. 21, 65 D.L.R. (2d) 353
[hereinafter *B.C. Offshore Reference*]

*Reference re the Ownership of the Bed of the Strait of Georgia and Related Areas*
see
*A.G. Canada* v. *A.G. British Columbia*

*Reference re the Seabed and Subsoil of the Continental Shelf Offshore Newfoundland*
(1984), 5 D.L.R. (4th) 384 (S.C.C.)
[hereinafter *Hibernia Reference*]

*Re Taxation Agreement Between Government of Saskatchewan and Government of Canada*
(1945), [1946] 1 W.W.R. 257 (Bd. of Arbitration)

# TABLE OF DEPARTMENTS, AGENCIES, AND BOARDS

## AUSTRALIA

Administrative Appeals Tribunal (AAT)

Australian Minerals and Energy Council (AMEC) and its Standing Committee on Offshore Petroleum Legislation

Commonwealth Bureau of Mineral Resources (BMR)

Commonwealth Department (In 1989 the relevant Commonwealth Department was called the Department of Primary Industries and Energy; because it has undergone a number of name changes over the past decade, it is hereinafter referred to simply as the Commonwealth Department.)

Commonwealth Department of Primary Industries and Energy

Commonwealth Department of Resources and Energy (DRE)

Commonwealth Parliament's Joint Committee of Public Accounts

Designated Authority (DA - state)

Joint Authority (JA - Commonwealth)

Joint Committee
    see
    Commonwealth Parliament's Joint Committee of Public Accounts

Senate Select Committee on Offshore Petroleum Resources

## CANADA

Canada, Department of Energy, Mines and Resources

Canada, Department of Indian Affairs and Northern Development

Canada-Newfoundland Offshore Petroleum Board (CNOPB)

Canada - Nova Scotia Offshore Oil and Gas Board

Canada - Nova Scotia Offshore Petroleum Board (CNSOPB)

Canada Oil and Gas Lands Administration (COGLA)

Canadian Coast Guard

Maritime Offshore Resources Board [proposed]

National Energy Board

Offshore Petroleum Standards Council [proposed]

Offshore Petroleum Training Standards Board [proposed]

Oil and Gas Committee (established under OGPCA, but disbanded)

Petroleum Directorate of the Newfoundland Energy Department

# TABLE OF STATUTES, BILLS, REGULATIONS, AND DIRECTIONS

## AUSTRALIA

*Administrative Appeals Tribunal Act 1975* (Cth) (No.91)

*Administrative Decisions (Judicial Review) Act 1977* (Cth) (No.59)

*Coastal Waters (State Powers) Act 1980* (Cth) (No.75)
[hereinafter *State Powers Act*]

*Coastal Waters (State Title) Act 1980* (Cth) (No.77)
[hereinafter *State Title Act*]

*Commonwealth Constitution Act*
see
*Commonwealth of Australia Constitution Act*

*Commonwealth of Australia Constitution Act*,
63-64 Vict., c.12 [U.K.]
[hereinafter *Commonwealth Constitution Act*]

*Environment Protection (Impact of Proposals) Act 1974* (Cth)
(No.164)

1985 Uniform Directions
see
*Specific Requirements as to Offshore Petroleum Exploration and Production - 1985*

*Petroleum (Ashmore and Cartier Islands) Act 1967* (Cth) (No.124)

*Petroleum Resource Rent Tax (Miscellaneous Provisions) Act 1987*
(Cth) (No.145)

*Petroleum Revenue Act 1985* (Cth) (No.188)

*Petroleum (Submerged Lands) Act 1967* (Cth) (No.118)
[hereinafter PSLA]

*Petroleum (Submerged Lands) Act 1967* (Vict.) (No.7591)

*Petroleum (Submerged Lands) Act 1967* (W.A.) (No.41)

*Petroleum (Submerged Lands) Act 1982* (Vict.) (No.9772)

*Petroleum (Submerged Lands) Act 1982* (W.A.) (No.33)

*Petroleum (Submerged Lands) Amendment Act 1980* (Cth) (No.80)

*Petroleum (Submerged Lands) Amendment Bill 1984* (Cth) (No.166)

*Petroleum (Submerged Lands) (Exploration Permit Fees) Act 1967* (Cth) (No.120)

*Petroleum (Submerged Lands) (Pipeline License Fees) Act 1967* (Cth) (No.122)

*Petroleum (Submerged Lands) (Production License Fees) Act 1967* (Cth) (No.121)

*Petroleum (Submerged Lands) (Registration Fees) Act 1967* (Cth) (No.123)

*Petroleum (Submerged Lands) Regulations* (Cth), Statutory Rules 1985, No.174 as am. by 1987, No.293

*Petroleum (Submerged Lands) (Royalty) Act 1967* (Cth) (No.119) [hereinafter *Royalty Act*]

*Royalty Act*
    see
    *Petroleum (Submerged Lands) (Royalty) Act 1967* (Cth)

*Seas and Submerged Lands Act 1973* (Cth) (No.161)

*Specific Requirements as to Offshore Petroleum Exploration and Production - 1985*, being a Schedule of the PSLA [hereinafter 1985 Uniform Directions]

*State Powers Act*

see

*Coastal Waters (State Powers) Act 1980* (Cth)

*State Title Act*

see

*Coastal Waters (State Title) Act 1980* (Cth) (No.77)

Territorial Sea and Continental Shelf Bill

## CANADA

*An Act to amend the Canada Shipping Act and to amend the Arctic Waters Pollution Prevention Act, the Maritime Code Act and the Oil and Gas Production and Conservation Act in consequence thereof*, S.C. 1987, c.7

*An Act to apply federal laws and provincial laws to offshore areas and to amend certain Acts in consequence thereof*, Bill C-104, 1st session, 33d Parl., 33-34-35 Eliz. II, 1984-85-86. [The Bill died on the order paper.]

*An Act to Protect Georges Bank*, S.N.S. 1988, c.56

*Atlantic Accord Act*

see

*Canada-Newfoundland Atlantic Accord Implementation Act*

*British North America Act, 1949* [U.K.]

see

*Newfoundland Act*

*Canada Labour Code*, R.S.C. 1985, c.L-2

*Canada-Newfoundland Atlantic Accord Implementation Act*, S.C. 1987, c.3 [hereinafter *Atlantic Accord Act*]

*The Canada-Newfoundland Atlantic Accord Implementation (Newfoundland) Act*, S.N. 1986, c.37 [hereinafter *Newfoundland Implementing Act*]

*Canada-Newfoundland Oil and Gas Spills and Debris Liability (Newfoundland) Regulations*, Nfld. Reg. 72/88

*Canada-Newfoundland Oil and Gas Spills and Debris Liability Regulations*, SOR/88-262 [hereinafter *Newfoundland Liability Regulations*]

*Canada - Nova Scotia Offshore Petroleum Resources Accord Implementation Act*, S.C. 1988, c.28 [hereinafter *Nova Scotia Accord Act*]

*Canada - Nova Scotia Offshore Petroleum Resources Accord Implementation (Nova Scotia) Act*, S.N.S. 1987, c.3 [hereinafter *Nova Scotia Implementation Act*]

*Canada - Nova Scotia Oil and Gas Agreement Act*, S.C. 1984, c.29

*Canada - Nova Scotia Oil and Gas Agreement (Nova Scotia) Act*, S.N.S. 1984, c.2

*Canada Oil and Gas Act*, S.C. 1980-81-82-83, c.81 [now R.S.C. 1985, c.O-6]

*Canada Petroleum Resources Act*, S.C. 1986, c.45 [hereinafter CPRA]

*Canada Shipping Act*, S.C. 1987, c.7

*Constitution Act, 1982*

*Frontier Land Registration Regulations*, SOR/88-230

*Income Tax Act*, R.S.C. 1952, c.148; R.S.C. 1970-71-72, c.63

*The Labour Standards Act*, S.N. 1977, c.52

*Newfoundland Act*,
R.S.C. 1970, Appendix II, No.30 (formerly *British North America Act, 1949* [U.K.])

*Newfoundland Implementing Act*
see
*The Canada-Newfoundland Atlantic Accord Implementation (Newfoundland) Act*

*Newfoundland Liability Regulations*
see
*Canada-Newfoundland Oil and Gas Spills and Debris Liability Regulations*

*Newfoundland Offshore Area Oil and Gas Operations Regulations*,
SOR/88-347 [hereinafter *Newfoundland Operations Regulations*]

*Newfoundland Offshore Area Registration Regulations*,
SOR/88-263 [hereinafter *Newfoundland Registration Regulations*]

*Newfoundland Offshore Area Registration Regulations*,
Nfld. Reg. 73/88

*Newfoundland Operations Regulations*
see
*Newfoundland Offshore Area Oil and Gas Operations Regulations*

*Newfoundland Registration Regulations*
see
*Newfoundland Offshore Area Registration Regulations*

*Nova Scotia Accord Act*
see
*Canada - Nova Scotia Offshore Petroleum Resources Accord Implementation Act*

*Nova Scotia Implementation Act*
see
*Canada - Nova Scotia Offshore Petroleum Resources Accord Implementation (Nova Scotia) Act*

*The Occupational Health and Safety Act*, S.N. 1978, c.23

*Offshore Petroleum Royalty Act*, S.N.S. 1987, c.9

*Oil and Gas Production and Conservation Act*, R.S.C. 1970, c.O-4 [hereinafter OGPCA]

*Oil and Gas Spills and Debris Liability Regulations*, SOR/87-331

*The Petroleum and Natural Gas Act*, R.S.N. 1970, c.294

*Terms of Union* [Newfoundland]
see
*Newfoundland Act*

*The Workers' Compensation Act*, S.N. 1983, c.48

# 1. INTRODUCTION

Canada's offshore petroleum resources are an asset that, over the long term, will play an important role in the country's energy security and competitiveness. These resources received considerable attention in the 1970s and early 1980s due to concerns about impending energy shortages. While the "energy glut" of the late 1980s reduced the interest of the industry in developing oil and natural gas reserves offshore Newfoundland and Nova Scotia, government incentives have begun to turn the situation around, as witnessed by announcements in 1988 concerning the Hibernia oilfield development.[1] Increasing investment in offshore projects will be accompanied by greater interest in the legal and administrative system that governs such projects. Thus, the time is ripe for a study of the complex arrangements that now govern Canada's east coast energy resources.

This study has been carried out in the context of a comparison between Canada's and Australia's offshore petroleum laws for several reasons. Both countries share a tradition of British common law. Both are federal states that have had difficulty sorting out the relationship between the two levels of government concerning offshore energy resources. In both countries, these disputes have been long-lasting, with resolutions at the negotiating table and in the courts. The objective of the study is to identify the similarities and differences in the resulting solutions and to ascertain whether lessons from one scheme have any useful application to the other.

The study begins with a description and analysis of the Canadian situation. Although British Columbia, the Yukon and the

---

1. *See* Canada, Dept. of Energy, Mines and Resources, *Hibernia Background Papers* (Ottawa: 1988).

Northwest Territories are mentioned, the Canadian section concentrates on Newfoundland and Nova Scotia, because the areas offshore these provinces have been the subject of the greatest amount of petroleum activity and, at the time the study was commenced, political Accords had been reached only between the federal government and the governments of those provinces.[2] An overview of the background to the Canadian regime is given, including the early federal-provincial negotiations, the *1982 Canada - Nova Scotia Agreement*, the failed Newfoundland negotiations that led to litigation, and the political changes in Canada that resulted in the 1985 *Atlantic Accord* and the 1986 *Nova Scotia Accord* with Newfoundland and Nova Scotia, respectively.

The legal regime for each of the two provinces is examined separately; the treatment of the two is the same except that the analysis of the Nova Scotia regime only emphasizes those features that differ from those of Newfoundland. Each section reviews the terms of the political Accord which provides the foundation for the legal regime and the legislative means by which the political Accords have been implemented. Attention is then devoted to the decision-making mechanisms for each regime. The division of legislative authority between the two levels of government is scrutinized, along with the fiscal arrangements and the procedures to be followed in resolving interprovincial boundary disputes.

This is followed by a look at the following selected legal and administrative issues:

- How can the Accords, their enabling statutes, and the resulting regulations be amended?
- To what extent is there or can there be consistency between the laws, regulations, policies, and administrative practices that pertain to these two areas and those that pertain to other

---

2. In September 1988, Agreements in Principle were reached between the federal government and the governments of the Northwest Territories and the Yukon. *See Northwest Territories Northern Accord Agreement in Principle* and *Yukon Northern Accord Agreement in Principle.* Negotiations toward a "British Columbia Accord" continue.

frontier areas?

- What are the relationships that exist between the Offshore Petroleum Boards, the energy departments of the two levels of government, and other federal and provincial departments and agencies?
- To what degree, if any, is there a lack of clarity concerning the legislative powers of the two levels of government? To what degree, if any, do the legislative powers of the two levels of government overlap?
- To what extent and in what way are decisions about offshore activities reviewable?

The study then turns to an analysis of the Australian system. The focus is only upon the two states (Victoria and Western Australia) where most of Australia's offshore petroleum exploration and production has taken place. The background to the current Australian arrangements is considered, including the *1967 Agreement*, the changing political climate that caused the federal Commonwealth government to pass legislation asserting jurisdiction over the offshore, and the states' unsuccessful challenge to the Commonwealth legislation in the High Court. The examination of the terms of the resulting *1979 Offshore Constitutional Settlement* parallels the Canadian section of the paper, considering the legislative implementation of the Settlement, the resulting decision-making mechanisms, the division of legislative authority between the two levels of government, the fiscal arrangements, and the means of setting and resolving disputes over boundaries. The Australian section concludes with an analysis of the same legal and administrative issues contained in the Canadian section.

The next part of the study compares the Canadian and Australian regimes, while the final section draws conclusions.

## 2. CANADA

### Background

The east coast offshore has been of interest to the petroleum industry since the early 1960s, when exploration programs were first undertaken. It soon became obvious that the areas were subject to the conflicting jurisdictional claims of the two levels of government. The *1964 Agreement*, reached between the Premiers of Nova Scotia, New Brunswick, Prince Edward Island, Newfoundland and Quebec, established offshore boundaries between these provinces, but there was no resolution at this time among the federal and provincial governments concerning offshore jurisdiction differences.[3]

The 1977 Memorandum of Understanding

One of the earliest formal attempts to settle federal-provincial offshore claims occurred on February 1, 1977, when the Prime Minister of Canada and the Premiers of Nova Scotia, New Brunswick, and Prince Edward Island signed a "Memorandum of Understanding in Respect of the Administration and Management of Mineral Resources Offshore of the Maritime Provinces" [hereinafter *1977 Understanding*].[4] One of the motivating forces behind the *1977 Understanding* was the complaint of oil industry representatives that continuing jurisdictional uncertainty would

---

3. Some background to the *1964 Agreement* is found in Jonathon I. Charney, "The Offshore Jurisdiction of the States of the United States and the Provinces of Canada — A Comparison" *in The Law of the Sea and Ocean Industry: New Opportunities and Restraints*, Proceedings: Law of the Sea Institute 16th Annual Conference, Halifax, N.S., 24 June 1982, at 427.
4. *See* Rowland J. Harrison, "The Offshore Mineral Resources Agreement in the Maritime Provinces" (1977-78) 4 Dalhousie L.J. 245.

make it difficult for production to proceed. The terms of the *1977 Understanding* deserve a brief review because of their contrast to the Accord regime that was put into place nearly a decade later and, as will be seen in the section entitled "Australia", because of their similarity to the state-federal arrangements in Australia.

The *1977 Understanding* covered the seabed and subsoil seaward from the ordinary low water-mark on the coasts of Nova Scotia, New Brunswick, and Prince Edward Island to the continental margin or to the limits of Canada's jurisdiction to explore and exploit the seabed and subsoil, whichever might be further. Interprovincially, the area had been divided according to the *1964 Agreement.* As between the federal and provincial governments, the offshore area was divided in two: the provinces would have jurisdiction over an area from the low water-mark to approximately five miles seaward, while the area beyond would come under the jurisdiction of a Maritime Offshore Resources Board. This Board was to have six members, three appointed by the federal government and one by each of the three provinces. The Board would carry out the functions of the federal energy minister as if the area were under exclusive federal jurisdiction and the functions of the provincial energy minister as though the area were under exclusive provincial jurisdiction.

The Board was to be funded on a 25/75 per cent basis by the federal and provincial governments, respectively, with offshore revenues being distributed on the same basis between the federal government and the province from whose offshore area the funds were generated. Provision was also made for the possibility of a revenue-pooling mechanism, to be negotiated between the provinces themselves. Sable Island was acknowledged to be within Nova Scotia, with one hundred per cent of revenues generated from that area going to the province.

Newfoundland did not participate in the *1977 Understanding*; continuing to assert its sole rights to the offshore area, its government viewed the Understanding as unfair and unacceptable. While the non-participation of Newfoundland would have undermined the effectiveness of the Understanding, as matters turned out, it was never implemented because a newly-elected government in Nova Scotia renounced further participation in the scheme. Without the involvement of Nova Scotia and Newfoundland, the provinces whose offshore areas were of principal interest to the industry, the *1977 Understanding* faded away and disputes about offshore jurisdiction continued.

## The 1982 Canada - Nova Scotia Agreement and the Canada-Newfoundland Negotiations

The *1982 Canada - Nova Scotia Agreement* [hereinafter *1982 Nova Scotia Agreement*] was reached to establish a framework for joint management of offshore petroleum. Doucet[5] has outlined some of the reasons why both governments were motivated to accept this compromise. He suggests that the federal government was anxious to promote offshore economic activity, the success of which could be attributed to its National Energy Program. It felt more confident about the strength of its legal position *vis-a-vis* Nova Scotia than it did *vis-a-vis* Newfoundland, and hoped that an agreement with Nova Scotia would pressure the Newfoundland government toward a compromise solution. For its part, Nova Scotia was unsure about the force of its legal arguments concerning offshore jurisdiction. Eager to promote economic activity of benefit to Nova Scotians, it feared that if Newfoundland reached a settlement first, industry dollars would be deflected because of the greater attractiveness of the Newfoundland offshore.

Within two years, the *1982 Nova Scotia Agreement* was enacted into law through a series of federal and provincial statutes.[6] The basic scheme underlying the arrangement was that the parties

---

5. G.J. Doucet, "Canada-Nova Scotia Offshore Agreement: One Year Later" (1984) 22 Alta. L. Rev. 132 at 132-133.
6. The federal enabling statute was the *Canada - Nova Scotia Oil and Gas Agreement Act*, S.C. 1984, c.29. Of the several provincial statutes, the most important was the *Canada - Nova Scotia Oil and Gas Agreement (Nova Scotia) Act*, S.N.S. 1984, c.2.

would put aside their jurisdictional differences and manage the offshore jointly through the Canada - Nova Scotia Offshore Oil and Gas Board on which the federal appointees would have three positions to the province's two, with the chairmanship *ex officio* residing in the Administrator of the Canada Oil and Gas Lands Administration [hereinafter COGLA]. All members of the Board would be federal or provincial civil servants.

The Board was assigned decision-making authority in relation to a number of matters, including the entering into of Exploration Agreements. In recognition of the federal majority position on the Board, however, the Nova Scotia members were given the right to exercise a "suspensive veto" in relation to certain matters. In effect, this enabled the Nova Scotia members to delay the implementation of decisions with which they disagreed. In fact, parties at both levels maintain that the Board members worked well together and that there were few disagreements over the several years of the Board's operation.

The *1982 Nova Scotia Agreement* provided for a $200 million Canada - Nova Scotia Development Fund, initially paid by the federal government, but repayable out of the province's royalty share beginning in the year after the first production from the Nova Scotia offshore. The province was entitled to receive all royalty revenues, rentals and bonus payments, and to levy provincial corporate and retail sales taxes. The offshore revenues, however, could not exceed the amount determined according to a complex formula set out in the *1982 Nova Scotia Agreement* which was designed to ensure that provincial revenues did not vastly exceed the nation's average per capita fiscal capacity. In other words, the provisions prevented the possibility of Nova Scotia becoming excessively rich relative to the rest of the country as a result of offshore petroleum development.

The *1982 Nova Scotia Agreement* also granted authority over Sable Island to the provincial Minister of Mines and Energy, giving the province the sole benefit of any revenues generated therefrom. Furthermore, it contained a "most-favoured province" provision; pursuant to clause 25, if the federal government entered into an offshore oil and gas management agreement with any other province prior to January 1, 1985, "the Nova Scotia government may substitute the latter agreement in its entirety for the present agreement in its entirety."

Whatever the federal government's hopes about the impact of the *1982 Nova Scotia Agreement* on its negotiations with Newfoundland, there was little positive effect. Newfoundland continued to maintain that it had brought the offshore with it when, in 1949, it joined Canada. Having lost its historical control over the exploitation of the fisheries at that time, the province felt that offshore petroleum resources should now be largely under its control. While revenues were a concern, the provincial position was influenced by fears that rapid development of the offshore would overheat the provincial economy, causing serious social impacts in relation to matters such as education and housing. Although Newfoundland had succeeded in having the offshore question put on the table as part of the negotiations concerning patriation of Canada's Constitution,[7] it had eventually joined eight of the other provinces in accepting the 1982 constitutional amendments even though they did not resolve the offshore issue.

Following patriation, negotiations with the federal government continued. By then, Newfoundland had seized upon the notion of joint management of the offshore as a result of an idea expressed by Prime Minister Trudeau in a speech in May 1981.[8] Despite this positive development, negotiations foundered and, in frustration, the Newfoundland government launched the *Newfoundland*

---

7. In 1982, major amendments to the Constitution were made which included, for the first time in Canada's history, a domestic amending formula. For a discussion of the process, *see* Roy J. Romanow, John D. Whyte & Howard A. Leeson, *Canada Notwithstanding* (Toronto: Carswell, 1984).
8. Prime Minister Pierre E. Trudeau, Address (Liberal Party Dinner, 5 May 1981) [unpublished]. The concept of "joint management" was touched upon only in passing reference in the speech. It is therefore something of an irony that this notion proved so crucial to the arrangements later worked out by the federal Conservative government.

*Reference*[9] to its Court of Appeal, the federal government countering with the *Hibernia Reference*[10] to the Supreme Court of Canada. After two decades of negotiations, it was now left to the courts to resolve the issue of offshore jurisdiction.

The Offshore Litigation

Before turning to the cases concerning the Newfoundland offshore, a word about earlier offshore litigation is required. In 1967, the *B.C. Offshore Reference* regarding jurisdiction over the British Columbia offshore had been decided by the Supreme Court of Canada.[11] The *Reference* dealt with the area beyond the territorial sea, as well as the area from the low water-mark to the territorial sea, and, on both points, the Court found jurisdiction to lie with the federal government. As will be seen in the section entitled "Australia", this decision may have had some impact upon the resolution of offshore disputes in Australia. The Newfoundland government never accepted the *B.C. Offshore Reference* as determinative of its position because Newfoundland had entered Confederation at a different time and pursuant to different arrangements than had British Columbia. In contrast, the federal Liberal government considered the *B.C. Offshore Reference* to be a powerful precedent in relation to Newfoundland.[12]

---

9. *Reference re Mineral and Other Natural Resources of the Continental Shelf* (1983), 145 D.L.R. (3d) 9 (Nfld. C.A.) [hereinafter *Newfoundland Reference*].
10. *Reference re the Seabed and Subsoil of the Continental Shelf Offshore Newfoundland* (1984), 5 D.L.R. (4th) 384 (S.C.C.) [hereinafter *Hibernia Reference*].
11. *Reference re Offshore Mineral Rights of British Columbia*, [1967] S.C.R. 792, 62 W.W.R. 21, 65 D.L.R. (2d) 353 [hereinafter *B.C. Offshore Reference*]. The *B.C. Offshore Reference* did not resolve contentious questions about jurisdiction over the area between mainland British Columbia and Vancouver Island. This issue was later resolved in the province's favour. *See A.G. Canada* v. *A.G. British Columbia*, [1984] 4 W.W.R. 289 (S.C.C.) (*sub nom. Reference re the Ownership of the Bed of the Strait of Georgia and Related Areas*).
12. This point of view is evident from Prime Minister Trudeau's speech, *supra*, note 8.

The Newfoundland Court of Appeal decided the *Newfoundland Reference* in February 1983.[13] The following question had been put to the Newfoundland Court of Appeal:

> Do the lands, mines, minerals, royalties or other rights, including the right to explore and exploit and the right to legislate, with respect to the mineral and other natural resources of the seabed and subsoil from the ordinary low-water mark of the Province of Newfoundland to the seaward limit of the continental shelf or any part thereof belong to or otherwise appertain to the Province of Newfoundland?

On the question of the continental shelf, the Court of Appeal held that jurisdiction is with the Government of Canada. The authority of the *B.C. Offshore Reference* led the Court to conclude that, as of 1871, Newfoundland's boundaries ended at the low water-mark. Although prior to joining Confederation in 1949 Newfoundland had the ability to exercise jurisdiction over the continental shelf, it had not done so. Thus, only the federal government could now be seen as having authority over this portion of the offshore.

On the other hand, the Court was able to find various acts of jurisdiction exercised by Newfoundland in relation to the territorial sea. The establishment of a Commission of Government in Newfoundland in 1934 had not detracted from the previous acts giving Newfoundland authority over the territorial sea nor, in the Court's view, did the 1949 *Terms of Union*[14] have a detrimental effect on the province's position. The Court pointed out, however, that the province's jurisdiction over a three-mile territorial sea[15] was "subject to any interference with these rights which might arise from any valid legislation of the Parliament of Canada in respect of the territorial sea and inland waters."[16]

---

13. *Newfoundland Reference*, *supra*, note 9.
14. The *Terms of Union* are a schedule to the *Newfoundland Act*, R.S.C. 1970, Appendix II, No.30 (formerly *British North America Act, 1949* [U.K.]).
15. The three-mile territorial sea reflects the state of international law at the time that Newfoundland exercised its authority over the area in question. Newfoundland could not be given the benefit of later international law developments that had the effect of enlarging the breadth of the territorial sea.
16. *Newfoundland Reference*, *supra*, note 9, at 41.

Over a year later, in March of 1984, the Supreme Court of Canada handed down its decision on the *Hibernia Reference*.[17] Notably, this *Reference* was narrower than the *Newfoundland Reference* launched by the Newfoundland government, asking whether Canada or Newfoundland had the following rights in relation only to an area around the Hibernia oilfield, in the continental shelf:

1) the right to explore the said mineral and other natural resources, and
2) legislative jurisdiction to make laws in relation to the exploration and exploitation of the said mineral and other natural resources.

In noting the Court of Appeal's decision in the *Newfoundland Reference*, the Supreme Court of Canada pointed out that that decision had been appealed by both governments: by the province as to the continental shelf and by the federal government as to the territorial sea.[18]

The Court accepted, for the sake of argument, that at least from 1926 to 1934 Newfoundland had had sufficient external authority to acquire continental shelf rights. Since such rights did not exist in international law at that time, however, the critical question was the position of Newfoundland from 1934 to 1949. The Court was of the view that the establishment of a Commission of Government in Newfoundland in 1934 meant that any continental shelf rights thereafter accrued to the Crown in right of the United Kingdom. On the assumption that continental shelf rights existed in international law by 1949, the Court was of the opinion that, on joining Canada, any such rights would have devolved to the only entity in Canada then possessing external sovereignty, namely, the Crown in right of Canada. In any event, though, the Court's review of international legal developments convinced it that, in 1949, international law did not as yet confer continental shelf rights. The conclusion that Canada has the rights to explore and exploit in the continental shelf led the Court easily to the view that Canada also has legislative jurisdiction, by virtue

---

17. *Hibernia Reference*, *supra*, note 10.
18. *Id.*, at 392.

of its residual "peace, order, and good government" power.

The outcome of the judicial decisions outlined above may be summarized as follows. Jurisdiction over the continental shelf off Canada rests with the federal government. This result has been reached specifically as regards British Columbia and Newfoundland, but it is widely conceded that the same result is inevitable in relation to other offshore areas, since Newfoundland's unique constitutional position provided the strongest possible argument for provincial continental shelf rights.[19] The territorial sea off British Columbia is also under federal jurisdiction, with the province having rights in relation to the waters lying between the mainland and Vancouver Island. The Newfoundland Court of Appeal's decision that Newfoundland has jurisdiction over a three-mile territorial sea cannot be considered final, despite the fact that the federal appeal on this point has been dormant for years. Even if the Court of Appeal decision concerning the territorial sea were upheld, Newfoundland's peculiar constitutional history might make such a ruling inapplicable to other offshore provinces such as Nova Scotia.

## Aftermath to the Litigation

The Supreme Court's decision in the *Hibernia Reference* was received bitterly by the Government of Newfoundland, which continued to maintain its moral entitlement to control the offshore. Its long history of high unemployment and its lack of control over other resources, such as the fishery, were undoubtedly factors that shored up its political resolve to continue the offshore battle despite its setback in the Supreme Court.

The federal Progressive Conservative Party had long taken the position that coastal provinces should play a significant role in

---

19. Rowland J. Harrison, "Jurisdiction Over Offshore Installations: Canada" *in* Ian Townsend Gault, ed., *Offshore Petroleum Installations Law and Financing: Canada and the United States* (London: International Bar Association, Section on Energy and Natural Resources Law, 1986) 39 at 56.

offshore management, a policy maintained during the party's short tenure in national office in 1979. A letter was sent to the coastal provinces by the then Prime Minister, The Honourable Joe Clark, contemplating a transfer of jurisdiction to the provinces, but the Conservatives were defeated before the matter could be fully resolved. Following the Supreme Court decision, the Newfoundland government noted the growing federal popularity of the Conservative party and commenced discussions with party representatives even though the Conservatives were still in Opposition. In these discussions, the province's earlier assertions of "ownership" were untenable due to the Supreme Court decision and the fact that such a position had been abandoned in the final stages of the ill-fated negotiations with the Liberal government.

During the 1984 federal election campaign, Brian Mulroney, the Conservative leader, expressed willingness to facilitate Newfoundland's participation in the offshore. An exchange of telegrams between Newfoundland's Premier Brian Peckford and Mr. Mulroney during the campaign formed the basis for the *Atlantic Accord*, discussed below, following the Conservatives' sweeping victory.

There are several possible reasons why the Conservatives were able to reach agreement with the Newfoundland government after so many years of failure by the federal Liberal government. As the analysis of the *Atlantic Accord* in the next section shows, the Conservatives were prepared to grant the provinces more rights in relation to the offshore than the Liberals had been. The Conservatives expressed the intention to reduce the scope of discretionary power that had been vested in the federal energy minister under the Liberal's National Energy Program: this policy approach may have made it easier for the Conservatives to accept the notion of delegating ministerial authority to an independent Offshore Petroleum Board, as required by the *Atlantic Accord*. Furthermore, during the course of the *Atlantic Accord* negotiations, world oil prices had declined, casting some doubt upon the economics of the Hibernia project. Finally, there is the important matter of bargaining power. By the 1984 election campaign, Newfoundland's legal position in relation to the offshore had been so weakened by the Supreme Court decision that the concessions offered by the Conservatives concerning revenue sharing and joint management must have appeared very attractive.

## The Newfoundland Regime

This section begins with an analysis of the *Atlantic Accord*, the political agreement entered into between Prime Minister Brian Mulroney and Newfoundland Premier Brian Peckford on February 11, 1985. The second part examines the federal and provincial statutes that implemented the Accord and considers some of the substantive changes that occurred in the process of transforming the Accord into legislation. The third part focusses on the decision-making regimes put into place by the legislation, emphasizing the interrelationships between the new Canada-Newfoundland Offshore Petroleum Board and the federal and provincial energy ministers. The final three parts discuss topics of particular interest to this study: the division of legislative authority between the federal and Newfoundland governments, the fiscal arrangements, and the provisions relating to the resolution of interprovincial offshore boundary disputes.

### The Atlantic Accord

The *Atlantic Accord* consists of 68 clauses, the opening clause providing that the Accord will be implemented "to the extent possible, through mutual and parallel legislation to be introduced by both governments into the Parliament of Canada and the Legislature of Newfoundland and Labrador." This language seems to reflect, at least in part, the legislative technique employed to implement the *1982 Nova Scotia Agreement*. There, through the device of "mirror legislation", each government had introduced legislation based upon the notion that it had jurisdiction over the offshore.[20] This method works well where there is constitutional uncertainty: the legal position of both parties is preserved while management decisions can be made in an agreed manner. In the Newfoundland situation, the Supreme Court of Canada's decision has removed uncertainty about continental shelf jurisdiction, but the question of the territorial sea remains in abeyance. The actual form that the "mutual and parallel legislation" has taken is discussed in the next part.

In clause 57, each party agreed to introduce implementing legislation within one year. This deadline proved to be an important factor in the process of transforming the *Atlantic Accord* into legislation.

Clause 68 of the *Atlantic Accord* provides that its terms apply to the area

> below the low water-mark lying off the coast of Newfoundland and Labrador out to the outer edge of the continental margin, coming within Canada's jurisdiction being north and east and south of the appropriate lines of demarcation between Newfoundland, the adjacent provinces, and the Northwest Territories.

While the Accord relates only to petroleum, clause 67 recognizes that discussions may be held in the future concerning the exploration, development, and production of other minerals.

Clauses 3 to 33 are concerned with the procedures for joint management of the offshore. It is agreed to establish the Canada-Newfoundland Offshore Petroleum Board [hereinafter CNOPB]. Clause 18 provides that the CNOPB's budget will be shared by the two governments, once it has been approved annually by each.

The *Atlantic Accord* makes provision for decisions to be made in one of four ways. Clause 24 empowers the CNOPB to make certain decisions without review, including the declaration of significant and commercial discoveries and the granting and

---

20. For a discussion of this technique, *see* Nigel D. Bankes, Constance D. Hunt & J. Owen Saunders, "Energy and Natural Resources: The Canadian Constitutional Framework" *in* Mark Krasnick, research co-ordinator, *Case Studies in the Division of Powers* (Toronto: University of Toronto Press, 1986) 53 at 87.

renewal of production licences. As a result of clauses 22, 23, 37 and 61, other matters are left to one of the two governments. In the federal sphere, these include Canadianization policy, legislation of general application and federal taxation, while provincial decisions include the royalty regime and other provincial-type revenue as well as provincial social and worker-protection legislation.

Beyond these three relatively clear demarcations of authority lies a more complex type of decision. "Fundamental decisions" include rights issuance, extraordinary powers (such as rights cancellation and drilling orders), and development plan approval. Final authority over the CNOPB's decisions concerning such matters rests with one of the ministers, depending upon whether the decisions relate to the pace and mode of exploration and the pace of production or to the mode of development, and whether Canada has achieved national self-sufficiency in energy and security of energy supply. If the party without the power to make a fundamental decision disagrees with the decision, clause 32 enables that party to suspend the execution of the decision for three months, to allow an opportunity for the achievement of consensus.

In relation to fiscal matters, clause 35 permits the provincial government to participate in pricing decisions in the same manner as other producing provinces and clause 37 entitles it to receive all royalties, provincial corporate income taxes, sales taxes, bonus payments, rentals and licence fees, and other general forms of resource revenue and to impose provincial taxes of general application. An Offshore Development Fund of $300 million is established, to be cost-shared on a 75/25 per cent basis between the federal and Newfoundland governments. A complex formula set out in clause 39 prevents Newfoundland's resource revenues from resulting in a dollar-for-dollar loss of equalization payments.

Clause 64 requires the federal government to introduce a mutually agreeable resolution into Parliament concerning the constitutional entrenchment of the *Atlantic Accord*, should the provincial government "achieve the requisite support among the other provinces". The exact legal nature of intergovernmental agreements in Canada being unclear,[21] such constitutional entrenchment would prevent a future federal government from amending unilaterally the political Accord. This clause appears to be based on the assumption that the Constitution's general amending formula would apply to these circumstances, requiring the agreement of the House of Commons, the Senate, and the Legislatures of two-thirds of the provinces having fifty per cent of the population. As Whyte has pointed out,[22] it is arguable that the appropriate amending provision is actually s.43 of the *Constitution Act, 1982*, which provides that a constitutional amendment concerning only one or more provinces may be made through agreement of the Senate, the House of Commons, and the Legislature(s) of the affected province(s).

Whatever the respective merit of these views, the matter is only of academic interest at present, since it is up to the Newfoundland government to take the initiative toward a constitutional amendment. No steps have as yet been taken to develop the necessary support from the other provinces, and so

---

21. See Bankes, *id.*, at 96-97, for a discussion of the law on this point. *See also* J. Owen Saunders, *Interjurisdictional Issues in Canadian Water Management* (Calgary: Canadian Institute of Resources Law, University of Calgary, 1988) at 82-120, where a distinction is drawn between "constitutionalized" and "non-constitutionalized" agreements, and, in the latter category, between those of a contractual nature and those of a political nature. The Macdonald Royal Commission also pointed to the unclear legal status of governmental agreements and recommended a constitutional amendment to establish a procedure that would enable Parliament and provincial Legislatures to enter into agreements that would bind successor governments. *See* Canada, Royal Commission on the Economic Union and Development Prospects for Canada, *Report*, vol.3 (Ottawa: Minister for Supply and Services, 1985) at 257-258 and 473. One example of an attempt to constitutionalize intergovernmental agreements is found in clause 3 of the schedule to the *Meech Lake Constitutional Accord*, which concerns agreements on immigration and aliens. For the text of the proposed provisions and a discussion of their effect, *see* Peter W. Hogg, *Meech Lake Constitutional Accord Annotated* (Toronto: Carswell, 1988) at 21-25.
22. John D. Whyte, "Issues in Canadian Federal-Provincial Cooperation" *in* J. Owen Saunders, ed., *Managing Natural Resources in a Federal State* (Calgary: Carswell, 1986) 322 at 327.

long as the federal Conservatives are in power, there seems little likelihood that the *Atlantic Accord* will be overturned.

## Legislative Implementation

After the *Atlantic Accord* was signed in February 1985, officials of the two governments worked together to draft the enabling statutes — the *Newfoundland Implementing Act* and the federal *Atlantic Accord Act.*[23] The provincial law was passed first, but both statutes were proclaimed on the same day. The two statutes are extremely similar, so for ease of reference the federal statute's provisions will be used hereinafter, unless otherwise indicated.

The process of drafting the enabling legislation was complicated by major changes to federal oil and gas policy that occurred at the same time. Under the previous Liberal regime, offshore oil and gas activities had been governed by the *Canada Oil and Gas Act.*[24] This statute had been a cornerstone of the National Energy Program, a policy that the Conservatives, in Opposition, had vowed to dismantle. A new federal energy policy for frontier areas, announced by the Conservative government in October 1985,[25] promised legislation to replace the *Canada Oil and Gas Act.* Clause 59 of the *Atlantic Accord* had anticipated that federal oil and gas legislation would be retained to the extent of its consistency with the Accord. Since federal officials saw the retention of "core" legislation as a means of ensuring common policies throughout the frontier areas despite bilateral arrangements with coastal provinces (and possibly with the northern territories),

---

23. *The Canada-Newfoundland Atlantic Accord Implementation (Newfoundland) Act*, S.N. 1986, c.37 [hereinafter *Newfoundland Implementing Act*] and *Canada-Newfoundland Atlantic Accord Implementation Act*, S.C. 1987, c.3 [hereinafter *Atlantic Accord Act*].
24. *Canada Oil and Gas Act*, S.C. 1980-81-82-83, c.81 [now R.S.C. 1985, c.O-6].
25. Canada, Dept. of Energy, Mines and Resources and Dept. of Indian and Northern Affairs, *Canada's Energy Frontiers: A Framework for Investment and Jobs* (Ottawa: 1985) at 7.

the legislation to implement the *Atlantic Accord* became inextricably tied to the new federal petroleum legislation.

The *Atlantic Accord Act* is divided into eight parts, several of which can be dealt with summarily for present purposes. Part VIII contains transitional, consequential and commencement provisions, while Part VII extends the application of federal tax laws to the Newfoundland offshore. Part VI establishes the Offshore Development Fund agreed to in the *Atlantic Accord*, spelling out pertinent administrative details such as the maximum level of the federal contribution, the means of establishing criteria for the selection of projects, the method of payment from the Fund, and the structure of a joint committee to oversee the Fund. Part V is concerned with the impact of the Accord on equalization payments.

Part II of the *Atlantic Accord Act* (Petroleum Resources) is largely modelled on the *Canada Petroleum Resources Act* [hereinafter CPRA],[26] while Part III (Production and Conservation) essentially tracks the provisions of the federal *Oil and Gas Production and Conservation Act* [hereinafter OGPCA].[27] The two Parts deviate from these federal petroleum statutes only to the extent necessary to implement the terms of the Accord. For example, certain powers that rest with the federal minister under the CPRA are exercisable by the Canada-Newfoundland Offshore Petroleum Board under the *Atlantic Accord Act*. Parts I (Joint Management) and IV (Revenue Sharing) will be examined in greater detail in the sections entitled "Decision-Making Mechanisms" and "Fiscal Arrangements", respectively.

The *Newfoundland Implementing Act* is similar to the federal *Atlantic Accord Act*, but lacks some of the parts contained in the latter (such as Parts IV (Revenue Sharing), V (Equalization Offset), and VII (Federal Corporate Income Tax)). The deviation

---

26. *Canada Petroleum Resources Act*, S.C. 1986, c.45 [hereinafter CPRA]. This is the statute intended to replace the *Canada Oil and Gas Act*, *supra*, note 24.
27. *Oil and Gas Production and Conservation Act*, R.S.C. 1970, c.O-4 [hereinafter OGPCA].

as to the latter two Parts can be explained by the fact that they concern matters within federal legislative jurisdiction. The revenue-sharing provisions appear only in the *Atlantic Accord Act* because, in clause 36 of the *Atlantic Accord*, the federal government had agreed to "incorporate by reference" those Newfoundland laws that would permit the provincial government to establish and collect resource revenues and provincial taxes of general application as if the petroleum-related activities were on land. Since Newfoundland laws on these subjects were already in place, there was no need for further provision in the *Newfoundland Implementing Act*. For a similar reason, only the federal s.152 refers to provincial "social" laws, reflecting the federal commitment in clause 61 of the *Atlantic Accord* to extend the application to the offshore of certain provincial laws. Moreover, the *Atlantic Accord Act* requires provincial consent to certain regulations, while the *Newfoundland Implementing Act* requires federal consent to the counterpart regulations; references in the *Atlantic Accord Act* to the federal minister are replaced in the *Newfoundland Implementing Act* with references to the provincial minister. Other minor differences between the two statutes resulted from changes made to the CPRA in committee or in the House of Commons after the *Newfoundland Implementing Act* had been passed, but most of these were eliminated when the *Newfoundland Implementing Act* was amended in 1987.[28]

There are some differences of substance between the *Atlantic Accord* and the *Atlantic Accord Act*, of which two examples follow.[29] First, clause 37 of the Accord had provided that Newfoundland would receive the revenue from offshore development. Section 212 of the Act, however, makes certain sums payable to the federal Receiver General, while s.214 provides

---

28. *Newfoundland Implementing Act*, *supra*, note 23.
29. It has been pointed out that lawyers did not control the drafting of the *Atlantic Accord*, but were assigned the task of translating it into legislation. Personal communication, Jan. 27, 1989, from Tamara Parschin-Rybkin of COGLA. This may help explain some of the differences between the Accord and the enabling legislation.

for the establishment of a Newfoundland Offshore Petroleum Resource Revenue Fund, the proceeds of which will be paid to the Newfoundland government. Thus, although Newfoundland will receive the monies as agreed in the Accord, it will be paid indirectly rather than directly (as contemplated by the Accord). This deviation was agreed to by the parties while the enabling statutes were being drafted because of legal views that, as a result of the Supreme Court decision, revenues generated from the offshore (at least from the continental shelf) were required to flow first to the federal treasury.

A second example concerns the arrangements for regional security in energy supply. Clause 54 of the Accord had provided that offshore hydrocarbons would be made available to the province on commercial terms "to meet both the total end use consumption and the feedstock requirements" of its industrial facilities. Section 41 of the *Atlantic Accord Act* appears somewhat more limited, giving Newfoundland a first option on petroleum only where there is a "shortfall of petroleum deliveries in the Province". This concession was accepted by Newfoundland because it recognized that its option would create commercial uncertainty for the industry and should, therefore, be narrowed in the legislation.

Section 2 of the *Atlantic Accord Act* gives legal status to the *Atlantic Accord* and any amendments thereto. While s.4 makes it clear that the *Atlantic Accord Act* and its regulations take precedence over any other Act of Parliament that applies federal or provincial law to offshore areas, there is no mention of the priority as between the Accord and the legislation. It would seem that the statute would take priority, since, as discussed earlier, the legal status of intergovernmental agreements is unclear. Since both the divergences described above were accepted by Newfoundland, it currently seems improbable that the issue of priority between the two documents will arise. It is possible, however, that such an issue could become important in the future: if the Accord itself were amended, a question could arise as to what arrangements govern in the period between the amendment of the Accord and the amendment of the enabling statutes.

Decision-Making Mechanisms

The *Atlantic Accord* was premised on the idea that the two governments would jointly manage offshore petroleum resources, largely through the establishment of an independent Offshore Petroleum Board. To a considerable extent, this Board would replace functions previously exercised by COGLA under the federal regime and by the Petroleum Directorate of the Newfoundland Energy Department under the provincial regime.

The CNOPB consists of seven members, three to be appointed federally and three provincially, with a jointly-appointed chairman.[30] So far the governments have agreed upon the selection of the first chairman as well as the other six members. Currently, the chairman and two vice-chairmen serve full-time and reside in Newfoundland, while some of the other members reside elsewhere. Section 11(1) of the *Atlantic Accord Act* provides that no members of the CNOPB may be civil servants, a provision that seems intended to keep the members somewhat distanced from government policy.

Section 17(2) gives the CNOPB a general advisory function in regard to the legislation and regulations. Beyond this, the starting-point of the *Atlantic Accord Act*, as set out in s.30, appears to be that the CNOPB's decisions are final, subject to the Act's provisions. In fact, many of the CNOPB's decisions are not final, but are subject to ministerial override, interaction with the Oil and Gas Committee (a review committee provided for in the legislation whose functions are described in the section entitled "Review of

30. The *Atlantic Accord Act*, *supra*, note 23, s.10. D. Paul Emond, in "A Critical Evaluation of the Atlantic Accord" (1985) 12 Resources, The Newsletter of the Canadian Institute of Resources Law, University of Calgary, has suggested that too much authority has been removed from governments by s.12 of the Act, which provides for a tie-breaking mechanism in the event that the two governments are unable to agree on selection of a chairman.

Decisions"), joint ministerial directives, or consultation with or approval of the federal and Newfoundland ministers. Because the CNOPB's powers are scattered throughout the Act, Table A has been drawn up to illustrate which of the CNOPB's decisions are actually final and which are subject to possible external control. As Table A - Part I reveals, among the matters over which the CNOPB has the final say are the prohibition of activities in the event of dangerous weather (s.56(1)(b)), the issuance of certain mandatory interests (s.57(2)), the selection of a bid in response to a call for bids (s.59(2)), the issuance of licences for subsurface storage of petroleum (s.86(1)), the issuance of operating licences and the authorization of work or activities in some circumstances (s.138(1)), and the approval of technical aspects of a development plan (s.139(4)(b)).

Table A - Part II lists matters designated as "fundamental decisions" [hereinafter FDs] that are governed by the procedure in ss.31 to 40. When the CNOPB makes a decision that falls under this rubric, it must notify both ministers, each of whom has thirty days in which to notify the CNOPB of its approval or disapproval of the decision. Silence is deemed to indicate approval, and, if both ministers accept the decision, either in writing or through silence, the CNOPB's decision may be implemented.

If the approval of both ministers is not obtained, the final say will rest with one of them, depending on the nature of the decision and other circumstances. The minister without the final say may suspend the other minister's decision for up to 90 days, after which the decision or amended decision will be implemented. The purpose of this arrangement is to enable the two ministers to attempt to reach agreement on the matter in dispute.

Which minister has the final say over what type of decisions, and under what circumstances? If national self-sufficiency and security of supply do not exist, the federal minister has the final say over all FDs *except* those pertaining to the approval of Part I of a development plan (which includes the production system to be used). Even in the latter situation, the federal minister may override the provincial minister's decision if it would unreasonably delay the attainment of self-sufficiency and security of supply. Section 34(4) provides that when self-sufficiency and security of supply exist, the provincial minister has authority over FDs. The Act sets out a procedure for determining when these conditions exist, although, under s.35(2), it is agreed that they do not exist up to December 31, 1990. Among the matters classified as FDs are the making of a call for bids (s.58(2)), the terms of significant discovery licences (s.73(4)), the making of drilling orders (s.76(1)), the attachment of terms and conditions to production licences (s.81(1)(b)), and the cancellation of interests in certain circumstances (s.123(2)).

All FDs are subject to the power of the two ministers to issue joint directives to the CNOPB, as provided by s.42. Other matters upon which joint directives may be issued are the approval of Canada-Newfoundland benefits plans and decisions to prohibit the commencement or continuation of work or activity in the case of dangerous or extreme weather conditions (see Table A - Part III).

Certain FDs, along with other matters, are also subject to the CNOPB's possible interaction with the Oil and Gas Committee. This matter is dealt with more fully in the section entitled "Review of Decisions"; for present purposes, it may be noted that the matters listed in Table A - Part IV are subject to the procedure set out in s.124 of the *Atlantic Accord Act*. When the CNOPB makes a decision subject to s.124 (including the issuance of a drilling order (s.76(1)) or the cancellation of an interest (s.79(3)), both of which are also FDs), the CNOPB must give the affected person thirty days advance notice of the anticipated decision. If requested to do so by the affected person, the CNOPB must remit the matter to the Oil and Gas Committee. Following the Committee's hearing, the CNOPB is required to consider the Committee's recommendation. A decision made pursuant to this process is also subject to judicial review (s.124(10)).

Table A - Part V lists a series of matters that further limit the CNOPB's freedom to act without the involvement of one or both ministers. For example, s.45(5) requires consultation with both ministers prior to the CNOPB approving a Canada-Newfoundland benefits plan, while an annual plan concerning calls for bids and issuance of interests requires the approval of the minister who, for the moment, has the fundamental decision-making power (s.43(1)).

Table A - Part VI sets out matters where the federal minister has a special role to play. As required by the *Atlantic Accord*, this includes decisions relating to Canadianization requirements for the granting of a production licence (s.89 *ff.*). The other special federal powers, set out in s.40, concern situations where there are supply shortfalls or energy emergencies.

It should be noted that, while many of the CNOPB's powers are circumscribed by other provisions in the Act, the CNOPB can be assigned additional powers through regulations. For example, pursuant to s.149(1)(j) and (k), the Governor in Council may make regulations authorizing the CNOPB to exercise powers relating to the removal of petroleum and concerning the design, construction, operation, or abandonment of pipelines. In this way, and through its enforcement functions, the CNOPB will have considerable powers in the day-to-day regulation of offshore petroleum.

As this review and Table A reveal, many of the CNOPB's most important decisions, as set out in the *Atlantic Accord Act*, are subject to political direction. The result is an incredibly complex web of interacting provisions that makes it a challenge to determine where the authority actually rests. This scheme makes it important for the CNOPB to be attuned to the political aspects of decisions within its purview; if it is not, its decisions may be overturned. Thus far, none of the CNOPB's FDs appear to have caused difficulty at the political level. The arrangements for the Hibernia project illustrate one method for dealing with the political dimension of certain matters, such as Canada-Newfoundland benefits.[31]

---

31. The Hibernia project owners have made certain undertakings in relation to industrial and employment benefits for Newfoundland and Canada. See *Hibernia Background Papers*, *supra*, note 1, at 12 *ff*. Through this method,

It remains to be seen what impact this complex decision -making arrangement will have on the industry's activities. So far, industry's day-to-day interactions with CNOPB staff on routine matters seem to have gone fairly smoothly, and operators tend to like the fact that the CNOPB is present in Newfoundland to provide a single window for regulatory purposes. But what will transpire if the CNOPB is faced with an FD where there is not necessarily the same view at both levels of government? Will the industry find it necessary to convey its point of view to the CNOPB as well as to the two governments? Such issues have not arisen yet in a significant way, but the experience in Australia, discussed later, may provide some insight into the future.

Legislative Authority

Another matter of interest is the way legislative authority, including the power to make regulations, has been divided between the two governments.

It should be noted that the CNOPB's authority in this regard is very limited. Section 23 empowers it to make by-laws in relation to such matters as the conduct of meetings, the manner of dealing with its business, and generally the management and carrying out of its work. Even this limited authority is subject to the Act and to the Accord.

Clauses 22 and 23 of the Accord state that each level of government has authority over certain classes of subjects. Federally, these matters include Canadianization policy, federal taxation, and decisions made under laws of general application. The *Atlantic Accord Act* is silent on the last point, probably because this power is presumed to exist by virtue of the *Hibernia Reference*, at least in regard to the continental shelf. Sections 234 and 235 extend the application of the *Income Tax Act*[32] to the

---

the two governments have agreed in advance what benefits arrangements are acceptable to them in the context of a major project.

32. *Income Tax Act*, R.S.C. 1952, c.148, now R.S.C. 1970-71-72, c.63.

offshore, while Canadianization policy is dealt with by ss.88 to 96. The latter provisions require that the holder of a production licence generally have a Canadian ownership rate of at least fifty per cent.

The *Atlantic Accord* gave Newfoundland authority over the royalty regime and other provincial-type revenues, and social legislation. Sections 206 to 217 of the *Atlantic Accord Act* make applicable to the offshore specified provincial statutes that deal with fiscal matters. This "incorporation by reference" of the provincial laws includes regulations made under them as well as any future amendments. Thus, should Newfoundland alter any of the listed statutes, the amendments would automatically apply to the offshore.

The term "social legislation" is defined in s.152(1) of the *Atlantic Accord Act*, and includes *The Labour Standards Act*,[33] *The Occupational Health and Safety Act*[34] and *The Workers' Compensation Act*,[35] all as amended from time to time. These laws and their regulations are made applicable, by virtue of s.152(2), to marine installations and structures that are in the offshore area "in connection with the exploration or drilling for or production, conservation or processing of petroleum". Complementing this provision is s.152(4), which removes the application of certain federal labour laws to installations and structures. Complicating s.152(2) slightly is s.152(3), which provides that the provisions of any of the social laws that relate to a matter respecting which certain regulations may be made under the *Atlantic Accord Act* do not apply to offshore structures and installations. This is discussed further in the section entitled "Legislative Authority".

While the Act authorizes the passage of regulations by the Governor in Council in relation to a myriad of topics, some of these powers are constrained by the procedure set out in s.7. This requires the federal minister to consult with the provincial minister before making certain types of regulations and provides that "no

33. *The Labour Standards Act*, S.N. 1977, c.52.
34. *The Occupational Health and Safety Act*, S.N. 1978, c.23.
35. *The Workers' Compensation Act*, S.N. 1983, c.48.

regulation shall be so made without the approval of the Provincial Minister". A complete list of the type of regulations to which this procedure applies is found in Table B.

Practically, it is anticipated that COGLA will continue to take much of the lead in drafting regulations, since, in relation to some subjects, COGLA is seen as having the best staff expertise. Moreover, the regulations will be important in harmonizing the regimes that apply to all offshore areas. COGLA's involvement will also be critical in steering new regulations through the federal regulatory process. While the legislation does not grant the CNOPB a role in relation to the design of regulations, it is recognized that the CNOPB ought to be consulted: as the day-to-day regulator of the offshore industry, it will often have experience useful to those who are drafting new regulations or amending old ones.

At least three sets of regulations have been made pursuant to the s.7 procedure.[36] The *Newfoundland Registration Regulations* were discussed in advance with the CNOPB and the Newfoundland and Labrador Petroleum Directorate, as well as with industry representatives and the federal departments of Energy, Mines and Resources and of Indian Affairs and Northern Development. The *Newfoundland Liability Regulations* were also discussed with the federal departments and industry and were approved by the province. The *Newfoundland Operations Regulations* were reviewed by the Canadian Petroleum Association and its affiliated organizations, the provinces of Newfoundland and Nova Scotia, and by the CNOPB. All these regulations are similar to the counterpart regulations that govern other frontier areas.[37]

---

36. *Canada-Newfoundland Oil and Gas Spills and Debris Liability Regulations*, SOR/88-262 [hereinafter *Newfoundland Liability Regulations*]; *Newfoundland Offshore Area Registration Regulations*, SOR/88-263 [hereinafter *Newfoundland Registration Regulations*]; and *Newfoundland Offshore Area Oil and Gas Operations Regulations*, SOR/88-347 [hereinafter *Newfoundland Operations Regulations*].
37. E.g., *Oil and Gas Spills and Debris Liability Regulations*, SOR/87-331 and *Frontier Land Registration Regulations*, SOR/88-230.

Since the provisions of the *Newfoundland Implementing Act* require the federal minister's consent to the making of certain provincial regulations, there will eventually be two complete sets of regulations pertaining to offshore management. Presumably, both sets will be identical, for all practical purposes.[38]

Fiscal Arrangements

Although the *Atlantic Accord* contained four measures to respond to Newfoundland's desire to receive substantial benefits from offshore petroleum resources, only three are reflected in the *Atlantic Accord Act.*

First, as already discussed, provincial revenue laws have been incorporated by reference in the *Atlantic Accord Act.* Without a "capping" provision of the kind that appeared in the *1982 Nova Scotia Agreement*, in theory the province will be free to set its royalties at whatever level it likes. This freedom is, however, subject to the reality that federal-provincial coordination will be necessary in the case of high-risk, expensive projects, a point demonstrated by the arrangements that have been made in regard to the Hibernia project.[39] There, the project sponsors submitted an initial fiscal proposal to both governments in October 1985, and, after nearly three years of negotiation, the three groups agreed to a set of principles to govern the project. For its part, Newfoundland has agreed to impose, prior to payout, a basic royalty of one per cent of gross revenues, to be increased by one per cent every eighteen months to a maximum of five per cent. This royalty will be reduced proportionately if current oil prices fall below U.S. $30, in 1987 dollars. After payout, the royalty will be the greater of five per cent of gross revenue and thirty per cent of net revenue. "Payout" will occur when cumulative revenues equal the sum of

38. This is the case, at least, in regard to the Newfoundland regulations made so far. *Canada-Newfoundland Oil and Gas Spills and Debris Liability (Newfoundland) Regulations*, Nfld. Reg. 72/88, and *Newfoundland Offshore Area Registration Regulations*, Nfld. Reg. 73/88.
39. *See Hibernia Background Papers*, *supra*, note 1.

eligible costs, pre-payout gross royalties, and a fifteen per cent return allowance. Once a before-tax return of eighteen per cent plus inflation has been attained by the sponsors, the province will impose a supplementary royalty of twelve and a half per cent of net revenues. Special arrangements have also been worked out in regard to the application of provincial retail sales and corporate income taxes. The federal government's prevailing taxes will apply and it will be entitled to a net profits interest of ten per cent of project net revenues, starting in the month after repayment of all guaranteed loans and any interest assistance.

Second, ss.228 to 233 of the *Atlantic Accord Act* provide for the establishment of the Offshore Development Fund. The federal contribution is limited to $225 million, and the province must provide one-quarter of the total costs incurred by both governments in respect of a project. Some see this arrangement as advantageous because it ensures that funded projects will be clearly linked to offshore activities. Criteria and procedures for the Fund may be determined by federal-provincial agreement, and provision is made for the establishment of a joint committee to oversee the Fund. In the context of the Hibernia project, the two governments have agreed to provide $95 million from the Fund towards the construction of facilities.

A third aspect of the fiscal arrangements, relating to equalization offset, is dealt with in ss.219 to 227 and is intended to ensure that the province does not suffer a dollar-for-dollar loss of equalization payments as a result of offshore revenues. Beginning in the first year of production, the federal government will pay the province an amount equal to ninety per cent of a year's reduction in equalization payments attributable to the increase in petroleum revenues. Beginning in the fifth year of production, the offset payment will be reduced by ten per cent for each subsequent year.

Although clause 35 of the *Atlantic Accord* states that the province will be entitled to participate in negotiations and consultations with the federal government "from time to time and in the same manner as the governments of other producing provinces for the establishment of the price of oil and natural gas in the offshore area", the implementing legislation is silent on this point.

Interprovincial Boundary Dispute Resolution

One issue that remains to be resolved is the demarcation of boundaries between the provinces, for example, as between Newfoundland and Nova Scotia. The *1964 Agreement* determined interprovincial boundaries between the five eastern provinces; the federal government, however, was not a party to this Agreement and, as a result, the issue is still open.

The *Atlantic Accord Act* does not define precisely the area which it covers. Section 2 defines offshore area as follows:

> those submarine areas lying seaward of the low water mark of the Province and extending, at any location, as far as
>
> (a) any prescribed line, or
>
> (b) where no line is prescribed at that location, the outer edge of the continental margin or a distance of two hundred nautical miles from the baselines from which the breadth of the territorial sea of Canada is measured, whichever is the greater.

While the Governor in Council is authorized by s.7 to make regulations for the purpose of prescribing lines to enclose areas adjacent to the province, provincial consent is required. However, s.6 empowers the federal minister to refer to arbitration any dispute between Newfoundland and another province that is also party to a joint management arrangement with the federal government, if the matter cannot be resolved by negotiation. The arbitrator is to apply principles of international law governing boundary limitations with such modifications as the circumstances require. Lines set as a result of such an arbitration do not require provincial consent.

No action has been taken as yet to set these lines. Although there are permits in areas that might be transversed by a boundary, no particularly promising discoveries have been made in such areas. As will be seen in the section entitled "Australia", however, the matter of interprovincial boundary demarcations may give rise to many other complicated issues.

## The Nova Scotia Legal Regime

The *1982 Nova Scotia Agreement* contained a "most-favoured province" clause.[40] Thus, when the *Atlantic Accord* with Newfoundland was signed in February of 1985, it was widely assumed that the *1982 Nova Scotia Agreement* would be renegotiated to give that province the benefit of an agreement as advantageous as the *Atlantic Accord*. In the event, this was what took place, notwithstanding that the most-favoured province clause of the *1982 Nova Scotia Agreement* had expired. The fact that the clause had required Nova Scotia to accept all of the provisions of another province's arrangement was apparently ignored.

### The Nova Scotia Accord

The description of the *Nova Scotia Accord* follows the same outline as that pertaining to the Newfoundland *Atlantic Accord*, except that here the emphasis is upon the aspects of the Nova Scotia arrangement that differ significantly from the Newfoundland regime.

The *Nova Scotia Accord* was signed in Halifax on August 26, 1986. While it replaces the *1982 Nova Scotia Agreement*, some of the ways in which it differs from the *Atlantic Accord* can be explained by the fact that it carries forward arrangements that had operated successfully under the *1982 Nova Scotia Agreement*. Other variations reflect legal, historical, and geographical differences between the two provinces. Between the 1970s and the 1980s, there has been a measure of competition between the two provinces for petroleum investment dollars, and, as a result, the Nova Scotia negotiators may have been reluctant to accept an agreement that was a carbon copy of the *Atlantic Accord*. Other differences may flow from the timing of the *Nova Scotia Accord*:

---

40. Clause 25 of the *1982 Nova Scotia Agreement*, the "most-favoured province" provision, allowed that, if the federal government entered into an offshore oil and gas management agreement with any other province prior to January 1, 1985, "the Nova Scotia government may substitute the latter agreement in its entirety for the present agreement in its entirety."

by the time it was negotiated, considerable experience had already been gained as a result of the process of negotiating the *Atlantic Accord* and converting it into legislation. It is not surprising, therefore, that one finds a slightly higher degree of detail in the *Nova Scotia Accord* than in the *Atlantic Accord.*

The preamble to the *Nova Scotia Accord* states that it will be implemented through "mirror" legislation, the device that had been used to implement the *1982 Nova Scotia Agreement.* This Accord describes the geographical area which it covers in great detail by adopting, through clause 43, the definition that had been used in the enabling legislation for the *1982 Nova Scotia Agreement.* A more detailed definition of geographical area was also required in the *Nova Scotia Accord* because it carries over from 1982 the special management arrangements for Sable Island. In contrast to the *Atlantic Accord*, there is no provision for the possibility that an offshore accord might be reached in the future as to other minerals, although of course the absence of such a provision in no way precludes the possibility of a future agreement as to other minerals.

The *Nova Scotia Accord* contains many sections concerning the establishment of a "unified" administrative regime. As of April 1989, the Canada - Nova Scotia Offshore Petroleum Board [hereinafter CNSOPB], which is to be set up according to the terms of the *Nova Scotia Accord*, has not yet been established. However, this paper describes the CNSOPB as though it is operational. As will be detailed in the section entitled "Decision-Making Mechanisms", the CNSOPB differs both in size and composition from the CNOPB and is subject to somewhat different decision-making arrangements. The *Nova Scotia Accord* distinguishes between matters that can be decided by the CNSOPB itself, those which are classified as "fundamental decisions" permitting ministerial interaction, and those in relation to which the ministers may give directives.

The fiscal provisions are similar to those in the *Atlantic Accord*, except that the equalization arrangments agreed to in the *1982 Nova Scotia Agreement* have been retained and the arrangements for the Canada - Nova Scotia Development Fund have been built upon the Fund that had been put into place in 1982. Like the *Atlantic Accord*, clause 42 of the *Nova Scotia Accord* contemplates the possibility of constitutional entrenchment.

Many involved in the administration of the offshore under the *1982 Nova Scotia Agreement* share the view that it worked well. One might ask, therefore, why the Nova Scotia government wished to enter into a new agreement. From a provincial point of view, the fiscal provisions of the *Nova Scotia Accord* (the lack of a "cap" on offshore revenue and the ability to set royalties) are seen to be an improvement over the 1982 Agreement. The *Nova Scotia Accord* also gave the provincal Legislature power over social legislation. Perhaps as importantly, the local presence of the CNSOPB in Halifax is seen to be an advantage. Although the provincial minister did not gain significant management powers under the *Nova Scotia Accord*, it is considered better to have decisions made by a Board which is half-appointed by the province than by a body ultimately controlled by federal authorities.

From a federal point of view, it was necessary to revisit the *1982 Nova Scotia Agreement* because of changes to the federal energy policy. The *Canada Oil and Gas Act* had reserved to the Crown a twenty-five per cent share of all oil and gas interests on federal lands. The 1982 Agreement had provided for the division of the Crown share between the two governments on lands administered under the Agreement. Because the Crown share was abolished as part of the federal Conservatives' new energy policy, the 1982 Agreement required revision.

## Legislative Implementation

Enabling statutes have now been passed by both governments, but have yet to be proclaimed.[41] Unless otherwise indicated, the

---

41. The federal statute is the *Canada - Nova Scotia Offshore Petroleum Resources Accord Implementation Act*, S.C. 1988, c.28 [hereinafter *Nova Scotia Accord Act*]. The provincial statute is the *Canada - Nova Scotia Offshore Petroleum Resources Accord Implementation (Nova Scotia) Act*,

statutory references are to the federal *Nova Scotia Accord Act.*

The scheme of the *Nova Scotia Accord Act* is similar to its Newfoundland counterpart. Part I deals with the joint management regime and the next two Parts are similar to the CPRA and the OGPCA, respectively, altered merely to reflect the provisions of the *Nova Scotia Accord.* The remaining Parts deal with Revenue Sharing, Equalization, the Canada - Nova Scotia Development Fund, a Drilling Fund, Crown Share Adjustment Payments, Corporate Income Tax, Transitional, and related matters. Two items not found in the *Atlantic Accord Act* (Part VII - Drilling Fund and Part VIII - Crown Share Adjustment Payments) are part of the "baggage" from the *1982 Nova Scotia Agreement* and will be explained in the section entitled "Fiscal Arrangements". The *Nova Scotia Implementation Act,* like its Newfoundland counterpart, lacks some of the Parts found in the federal statute because they are unnecessary or inappropriate in a provincial statute.

As with the Newfoundland enabling legislation, there are some differences between the federal and provincial statutes that reflect the fact that they are the instruments of different levels of government (for example, reference to a provincial rather than a federal minister and to the Lieutenant Governor in Council rather than to the Governor in Council).

The *Nova Scotia Implementation Act* also contains some sections for which there are no counterparts in the federal *Nova Scotia Accord Act.* For example, s.53 provides:

> The provincial Minister and any corporation, agency, board, commission or tribunal of the Province may exercise such powers, duties and functions as are conferred on the Provincial Minister or the corporation, agency, board, commission or tribunal pursuant to the federal *Implementation Act.*

This section has been included in the provincial statute because of the view of Nova Scotia's lawyers (a view not shared by the federal lawyers) that mere federal incorporation by reference of provincial laws does not automatically amount to the adoption of

---

S.N.S. 1987, c.3 [hereinafter *Nova Scotia Implementation Act*].

the provincial administrative system.

Another difference between the two pieces of legislation deserving comment is the definition of "Nova Scotia lands" found in s.2(p) of the provincial Act but nowhere in the federal *Nova Scotia Accord Act.* It defines such lands as Sable Island and "those submarine areas that belong to Her Majesty in right of the Province or in respect of which Her Majesty in right of the Province has the right to dispose of or exploit the natural resources, and that are within the offshore area". Section 8 of the federal Act makes the Act applicable to Nova Scotia lands within the offshore area (the latter being defined in detail in a Schedule to the Act). It will be recalled that there has been no judicial determination of jurisdiction specifically in relation to the Nova Scotia offshore; the province continues to maintain that it has jurisdiction over a large area[42] and these specific references are consistent with the province's legal position.

## Decision-Making Mechanisms

As of April 1989, the CNSOPB had not been established. In the interim, COGLA's Halifax office is playing a bridging role between the decision-making arrangements from the *1982 Nova Scotia Agreement* and those under the *Nova Scotia Accord.* To be more specific, both governments are attempting to follow the spirit of the Accord even though, from a legal point of view, the legislation that implemented the *1982 Nova Scotia Agreement* continues to govern. The discussion that follows, then, is not based on actual experience, but rather on an analysis of the arrangements that will be put into place.

The CNSOPB will have only five members, two appointed by each government, with the chairman jointly appointed.[43] Some of those negotiating the *Nova Scotia Accord* felt that a seven member

---

42. For a description of the basis of Nova Scotia's legal claim, *see* Edward C. Foley, "Nova Scotia's Case for Coastal and Offshore Resources" (1983) 13 Ottawa L. Rev. 281.
43. *Nova Scotia Accord Act, supra,* note 41, s.10.

Board (such as in Newfoundland) would be unnecessarily large, particularly in view of the relatively slow pace of offshore activity in the late 1980s. If it is later deemed necessary, the size of the CNSOPB can be expanded. Not more than one of each government's appointees may be a member of the civil service.[44] It is widely assumed that each government will appoint one member from its civil service. In some quarters, this is seen as advantageous because civil service members may help bring a government point of view to CNSOPB deliberations. On the other hand, some observers suggest that such appointees may not necessarily be in a position to communicate the views of, for example, the minister's office. Another reason sometimes given for the possible use of civil servants on the CNSOPB is economy: until there is more to occupy the CNSOPB's attention, at least two of the CNSOPB's members can be gainfully employed at their normal work. As some have pointed out, only three members of the CNOPB are currently employed full-time on CNOPB activities.

Like the Newfoundland arrangement, the Nova Scotia enabling legislation grants the CNSOPB the power to make the final decision in relation to some matters while, in relation to others, there is a degree of political control or the necessity to interact with the Oil and Gas Committee. As illustrated by Table C, the list of matters falling into the various categories is virtually identical to that for the CNOPB. Some of the main differences are highlighted below.

The list of matters classified as "fundamental decisions" [hereinafter FDs] under the two regimes is very similar except that in Nova Scotia, the issuance of a licence for the subsurface storage of petroleum is treated as an FD (s.89), while in Newfoundland it is a matter for the CNOPB alone (s.86). The procedure for handling FDs, set out in ss.32 to 37 of the *Nova Scotia Accord Act*, is different than that for Newfoundland. In Nova Scotia, the CNSOPB is required to notify each minister of any FD. Within

---

44. *Id.*, s.11.

thirty days, either minister may notify the CNSOPB of its intention to suspend the FD for a period of up to sixty days. It is also possible, within the thirty-day period and the suspension period, for the ministers to jointly set aside the FD, or for the provincial minister to set aside an FD that relates to Part I of a development plan or to a call for bids or interests relating to Sable Island or the Bay of Fundy (as discussed below).

If the federal minister is of the view that the FD or the setting aside of an FD by the provincial minister would unreasonably delay the attainment of security of supply, he may set aside the FD himself or overrule the setting aside of an FD by the provincial minister. If the provincial minister disagrees with the federal minister's decision in this regard, he may in effect launch an appeal to the National Energy Board, whose determination of the issue (in accordance with various definitions and formulae set out in the federal Act) is final. One reason for this somewhat convoluted arrangement for making decisions may be that, since natural gas rather than oil is likely to be produced offshore Nova Scotia, it would be difficult to make a case that a decision is interfering with the attainment of national energy security. On this reasoning, the federal veto power is quite narrow in scope. It should also be noted that, in contrast to Newfoundland, the general arrangements for FDs under the *Nova Scotia Accord Act* do not contemplate the provincial minister ever having the main authority over FDs, except as regards Sable Island and the Bay of Fundy.

A comparison between Part III of Tables A and C reveals a different approach to the matter of ministerial directives. Like the Newfoundland arrangement, s.41 of the *Nova Scotia Accord Act* contemplates joint ministerial directives being given to the CNSOPB on several matters, including all FDs, Canada - Nova Scotia benefits plans, public reviews, studies, and advice with respect to policy issues. The Newfoundland joint ministerial directive power, in s.42 of the *Atlantic Accord Act*, is broader in that it includes the CNOPB's prohibition of work or activities due to extreme weather conditions. Under s.43 of the *Nova Scotia*

*Accord Act*, the CNSOPB is obligated to submit an annual plan to the ministers outlining its anticipated decisions in relation to calls for bids, issuance of interests, and exploration and development. Pursuant to s.43 of the *Atlantic Accord Act*, the minister with the FD authority may reject the plan. A similar result is achieved in Nova Scotia in a different way through s.41(2), which entitles either minister to issue a directive to the CNSOPB to specify what parts of the offshore should be included in a call for bids. Another difference between the two is that, in Nova Scotia, s.41(3) empowers the federal minister to issue a directive to the CNSOPB respecting the Canadian ownership rate in relation to any interest.

The legislation contains special arrangements for Sable Island and the Bay of Fundy, the former area having also received special treatment under the *1982 Nova Scotia Agreement*. There are two provisions of note. First, s.41(4) entitles the provincial minister to issue the CNSOPB with binding directives with respect to "any fundamental decision relating to the Bay of Fundy or Sable Island." Second, s.35(1)(b)(ii) empowers the provincial minister to set aside "a fundamental decision with respect to a call for bids pursuant to Part II in relation to, or interests in relation to, a portion of the offshore area that is situated wholly within the Bay of Fundy or Sable Island."

Legislative Authority

The Nova Scotia and Newfoundland arrangements differ slightly as regards the applicability of provincial social and revenue legislation. Specifically, the lists of applicable provincial laws are a little different, perhaps because Nova Scotia's legislation is differently organized than Newfoundland's or because the province's "wish list" was not identical.[45] As a comparison of Tables B and D will reveal, the Nova Scotia list of regulations to be jointly-agreed is somewhat longer than

45. The relevant sections for Newfoundland are 152 and 206, and for Nova Scotia, 157 and 211.

Newfoundland's. In other respects, the arrangements for legislative authority are much the same.

Fiscal Arrangements

Royalties are reserved to the Crown in right of Canada pursuant to s.99 of the *Nova Scotia Accord Act*, but the rate is left to the province through a federal adoption of the province's *Offshore Petroleum Royalty Act.*[46] Although a royalty rate had not been set under this legislation as of April 1989, the province has released a discussion paper outlining its ideas on the subject.[47] Part IV of the *Nova Scotia Accord Act* provides for the incorporation by reference of a series of other provincial revenue-raising statutes, while Part V concerns the equalization offset payments.

The arrangements for the Canada - Nova Scotia Development Fund are contained in Part VI, which continues the existence of the Fund set up under the *1982 Nova Scotia Agreement's* enabling legislation. Section 237 limits federal payments to $200 million, including any amounts paid under 1982 Agreement.[48] Unlike the 1982 Agreement, this money takes the form of a federal grant rather than a loan. Pursuant to Part VII of the *Nova Scotia Accord Act*, an additional amount of up to $25 million may be paid to Nova Scotia Resources (Ventures) Limited, in respect of its Canadian exploration expenses and Canadian development expenses that could contribute to early development of petroleum in the offshore area. Section 240(2) sets out a formula concerning

---

46. *Offshore Petroleum Royalty Act*, S.N.S. 1987, c.9.
47. Nova Scotia, Dept. of Mines and Energy, *News Release* (n.d.). For a discussion, *see* S.J. Snook, "Petroleum Royalties: Nova Scotia's Proposed Royalty Regime and Comparisons to Norway, United Kingdom and Australia", Paper No.87-38-96 (Petroleum Society of CIM).
48. Payments made under the Canada - Nova Scotia Development Fund have been criticized by Canada's Auditor General. In his 1988 report, he noted that more than $100 million had been allocated to projects, including highways and bridges, "which, because of their geographical location, would require a leap of imagination to visualize being linked to offshore development." "Highlights" *Toronto Globe and Mail* (14 December 1988) 1.

the eligibility of such expenses, which must be "reasonable in the circumstances" and relate to activities in the period between August 27, 1986, and December 31, 1991.

Part VIII of the *Nova Scotia Accord Act* is entitled "Crown Share Adjustment Payments" and is meant to compensate the province for revenue that it would have received under the *1982 Nova Scotia Agreement* by virtue of its entitlement to half the twenty-five per cent Crown share. The complex procedures for determining the amount of such payments are contained in s.247, with the onus on the provincial minister to convince the federal minister that, had the province been entitled to acquire a portion of the Crown share in a field under the earlier arrangement, "the rate of return in respect of that project that would have been obtained on behalf of Her Majesty in right of the Province, calculated in the manner prescribed, would have been equal to or greater than the threshold rate of return." The term "threshold rate of return" is defined in s.246(1).

Another carry-over from the *1982 Nova Scotia Agreement* that may have financial implications is s.40. Like the earlier Agreement, this guarantees the province the right to obtain up to fifty per cent ownership in a trunkline for the transmission of petroleum from the offshore area to the province.

### Interprovincial Boundary Dispute Resolution

Like the Newfoundland arrangement, s.5(1) of the *Nova Scotia Accord Act* authorizes the Governor in Council, with the provincial minister's consent, to make regulations amending the Schedule which defined the "offshore area", while s.48 permits the settlement of interprovincial boundary disputes by binding arbitration.

## Selected Legal and Administrative Issues

The complexity of the arrangements that now govern offshore petroleum management on Canada's east coast give rise to a number of interesting legal and administrative issues. Those selected for examination below are not meant to be exhaustive, but are intended to provide a counterpoint for the analysis of the Australian regime.

## Amendment of the Accords, the Enabling Legislation, and the Regulations

Both the *Atlantic Accord* and the *Nova Scotia Accord* refer to the matter of amendment. Clause 2(g) of the *Atlantic Accord* states that one purpose of the Accord is to provide for a stable and permanent management arrangement "by providing that the Accord may only be amended by the mutual consent of both governments", while clause 60 provides: "Except by mutual consent, neither government will introduce amendments to the legislation or regulations implementing the Accord." Clause 1.03 of the *Nova Scotia Accord* requires the parties to review the objectives of the Accord every five years or whenever requested by either party, stating that the objectives may be amended at any time by mutual agreement. Clause 39.08 further provides that, subject to a few exceptions, "neither the Governor General in Council nor the Lieutenant Governor in Council shall make, repeal or amend orders or regulations made pursuant to the legislation implementing the Accord."

The enabling legislation for Newfoundland does not deal with amendments to the *Atlantic Accord*, but the preamble states that the two governments have agreed that neither will introduce amendments to the Act or any regulations made thereunder without the consent of both. Section 7 lists the kind of regulations that are subject to the joint amendment process. Section 6 of the *Nova Scotia Accord Act* also deals with jointly-agreed regulations, while s.7 states that the two governments may jointly amend the Accord from time to time. This statute is silent as to the procedure to be followed in amending the statute itself.

What is the effect of these provisions? From a political perspective, there is a strong compulsion on all three governments to undertake amendments only with the consent of the other level. So long as the regimes are working well and the governments do not change their policies, there is no reason to expect any deviation from the agreed process. From a legal point of view, s.6 of the *Nova Scotia Accord Act* and s.7 of the *Atlantic Accord Act* require federal-provincial agreement upon the specified regulations; while these provisions remain in the statutes, there must be compliance with the stated procedure.

Is there anything to prevent the federal government, however, from amending s.6 and s.7, or from altering unilaterally the overall scheme provided for in these statutes? From a legal point of view, the answer is probably "no": the uncertain status in Canadian law of intergovernmental agreements means that the provisions of the Accords are unlikely, in law, to prevent the federal government from amending the enabling statutes without consulting the two provinces. *Vis-a-vis* the continental shelf off Newfoundland (and, arguably, off Nova Scotia), the *Hibernia Reference* gives the federal government the power to do so. The issue of jurisdiction over the territorial sea off Newfoundland is unresolved in the face of the dormant appeal of the *Newfoundland Reference*; a federal decision to alter either regime on its own would undoubtedly bring the legal issues to the fore again.

It is precisely for these reasons that provision has been made in the Accords for constitutional entrenchment: because of the long history of acrimony over offshore jurisdiction, the provinces wished the security that only constitutional entrenchment can provide. One wonders, therefore, why Nova Scotia and Newfoundland have taken no steps to seek support from the other provinces on the matter of entrenchment. One reason may be that, so far, the Accords are working harmoniously. In the one situation where a federal-provincial rift loomed on the horizon, a potential clash of views was averted.[49] The longer the Accords are in place,

---

49. This concerned drilling on Georges Bank, off Nova Scotia; it is discussed in the section entitled "Relationships Between the Offshore Petroleum Boards, the Federal and Provincial Governments, and the Federal and

operating satisfactorily, the less the likelihood that a new government will alter them. Another reason is probably that, in the period since the Accords were negotiated, the national and intergovernmental agenda has been fully occupied with the pressing issues of the Canada-U.S. free trade agreement and the *Meech Lake Accord* (which raises many other constitutional issues). Finally, Newfoundland and Nova Scotia may consider it prudent to await the result of negotiations concerning the "British Columbia Accord":[50] with British Columbia onside, the impetus toward constitutional entrenchment would be strengthened.

Assuming that all governments will comply with the agreed amending procedures, what implications does this have for federal and provincial energy policies? It is suggested that procuring amendments to the Accord statutes may prove difficult. The federal government is committed to having a common core of legislation for the offshore; this will be accomplished by keeping Parts II and III of the Accord statutes in line with the CPRA and the OGPCA and having regulations with relatively common provisions for all areas.

If the need for a statutory amendment is identified, then the federal interest will be to ensure that the counterpart provisions for other areas receive the same treatment. This will require revision of three federal (CPRA or OGPCA and the two Accord statutes) and two provincial Acts (the Newfoundland and Nova Scotia implementing statutes). The situation will grow even more complex as a result of the 1988 agreements with the Yukon and the Northwest Territories governments and a possible future agreement with British Columbia.[51] The degree of consultation and

---

Provincial Agencies".

50. "British Columbia Accord", *supra*, note 2.
51. *Supra*, note 2. The *Northwest Territories Northern Accord Agreement in Principle* contemplates that the territorial governments will have legislative jurisdiction over their onshore petroleum regimes, which may result in some diversity. Joint regimes with the federal government, however, are contemplated for the offshore, enabling the federal government to promote uniformity in the overall offshore regime but requiring the participation of the territorial governments in any legislative amending process.

coordination required will make the amending process slow and cumbersome, even when the parties agree on the needed amendments. New federal proposals will have to be sold to the offshore provinces, making it difficult for the federal government to initiate major policy changes that do not find favour with one or more of the offshore provinces. Given the relatively untested state of the Accord legislation, this could result in rigidity that might prove undesirable as experience dictates the need for legislative changes.

On the other hand, it is noteworthy that the CPRA was developed through what was essentially consensus between the Newfoundland, Nova Scotia, and federal governments. Indeed, it is arguable that the system may actually be strengthened by the joint amendment process: with more groups thinking about the same problems, more imaginative solutions may result.

A major test of the amending process will be the overhaul of the OGPCA, being discussed among COGLA, the industry, the CNOPB, and representatives of the governments of Nova Scotia, British Columbia, and the two territories. Although there is concern in some quarters that these changes will take a long time, others believe that, with a spirit of cooperation, changes can be procured expeditiously. A great deal of the success or otherwise will rest on the shoulders of COGLA, whose role is changing from day-to-day regulator to coordinator.

## Consistency in Laws, Regulations, and Policies

The goal of consistency between the regimes governing different offshore areas was mentioned in the objectives sections of both Accords, although each provision contained the limiting language "insofar as is appropriate". This section considers what mechanisms are available to achieve this goal and the extent to which consistency may be attainable in the future.

***Mechanisms for Consultation***

The bilateral system for offshore petroleum management that has developed in Canada requires the involvement of individuals in several institutions, including the two levels of government and the CNOPB and the CNSOPB. The *Regulatory Recommendations* of the federal energy minister's Task Force on *Ocean Ranger* underscored the need for maximum consistency of the requirements and standards in all federal and provincial administrations, and emphasized the desirability of a single forum to bring together federal and provincial regulators to coordinate their procedures.[52] To accomplish this, the Task Force recommended the establishment of an Offshore Petroleum Standards Council, comprised of the chairman of each regional Offshore Petroleum Board and the head of COGLA.[53] This Council could designate inspectors for all jurisdictions. As well, the Task Force proposed the establishment of a statutory board, the Offshore Petroleum Training Standards Board, to develop training standards and personnel qualifications. This body would be made up of representatives of the federal-provincial Offshore Petroleum Boards, COGLA, the Canadian Coast Guard, industry, educators, and employees. Its standards and assessments could be overruled by the Council. With the exception of the power to appoint inspectors and to overrule the decisions of the Training Standards Board, the Council's decisions would be implemented by the CNOPB, the CNSOPB, and COGLA, thus preserving the jurisdiction of each regional authority.

---

52. *See* Canada, Minister of Energy, Mines and Resources Task Force on *Ocean Ranger*, *Regulatory Recommendations — The Promotion and Enhancement of Safety in Oil and Gas Operations on Canadian Frontier Lands; Equipment Procedures Personnel* (Ottawa: 1986) [hereinafter *Harrison Task Force Report*] at 83-84. The Task Force was appointed to advise the minister on certain recommendations that had been contained in Part II of the *Report* of the Royal Commission on the *Ocean Ranger* Marine Disaster, notably a proposal on the form and substance of omnibus legislation specific to Mobile Offshore Drilling Units [hereinafter MODUS] and a related proposal to ensure consistency in the regulation of offshore oil and gas operations.
53. *Id.*, at 85.

These recommendations have not yet been implemented, but are being addressed in the context of the review of the OGPCA. In general, the above recommendations seem to have been favourably received by provincial authorities and the CNOPB, although there is some difference of opinion concerning the need to establish, through legislation, the proposed Training Standards Board.

The matter of intergovernmental and interagency consultation will be considered further in the section entitled "Relationships".

***Uniformity in Petroleum Laws and Administration***

As has been seen, offshore petroleum management on the east coast is now governed by both federal and provincial statutes.[54] The federal government's control over its statutes and regulations, and the influence it can bring to bear on the two provinces, will help to ensure that desired consistency is maintained in laws and regulations. Because the federal government has incorporated by reference provincial laws on matters such as royalties, it is inevitable that there will be differences in this regard between the two areas.

Since the Offshore Petroleum Boards will play a major role in day-to-day regulation, the industry might face different policies and administrative practices depending upon whether it is operating offshore Newfoundland, Nova Scotia, or in the Arctic. To some extent, this concern has already begun to be addressed. The CNOPB, COGLA, and Nova Scotia officials have worked together to develop a common set of documents to be used in relation to rights issuance in all frontier areas. As a result of this process, the CNOPB circulated draft copies of its Exploration Licence, Significant Discovery Licence, and Production Licence for industry comment in October 1987. The cooperation that

---

54. As of April 1989, the Nova Scotia offshore is governed by the legislation enabling the *1982 Nova Scotia Agreement*. It will eventually be governed by the federal *Nova Scotia Accord Act* and the provincial *Nova Scotia Implementation Act*. The Newfoundland offshore is governed by the federal *Atlantic Accord Act* and the provincial *Newfoundland Implementing Act*.

characterized the development of these forms is seen as an example of how other matters may be addressed in the future. The three groups have also achieved some consensus on the means whereby oil companies can establish proof of financial responsibility for liability arising from spills and debris.

Some differences among the areas may be a necessary reflection of local circumstances. For example, the geography and climate offshore Newfoundland may make it sensible to have only one call for bids each year there, even if procedures are different elsewhere. The contrasting rental refund policies that have been developed for Newfoundland and Nova Scotia reflect differing views of the economics of offshore exploration. Other deviations, such as the way in which significant discovery areas are delineated, may be a result of vague definitions in the legislation that can give rise to varying interpretations.

It is also anticipated that there may be minor deviations in the occupational health and safety regulations that will apply on the east coast and those that will apply to non-Accord frontier areas. This is because the enabling authority for the latter areas comes from Part IV of the *Canada Labour Code*, while for the Accord areas the authority flows from the enabling statutes. The fact that these enabling sections are not identical to those in the federal *Canada Labour Code* may result in some differences in the way the regulations have to be drafted.

When there are deviations in the Accord statutes and administrative practices among the various areas, what is the impact upon industry? On the one hand, there is broad agreement that the Accord regimes are a vast improvement over the days when the two levels of government were engaged in serious disputes over offshore jurisdiction. The petroleum industry has coped with different provincial regimes in western Canada and should be able to cope with a similar situation in frontier areas. On the other hand, the complexity of developing frontier regimes may impose certain costs. For example, considerable time will have to be spent ensuring that personnel, such as industry accountants, are fully apprised of the minor differences that pertain to the various areas. Preparation of legal opinions will be complicated and made more time-consuming by the fact that the many statutes and regulations to be cited are similar, but not necessarily identical.

### *Uniformity in Other Laws*

Pursuant to the offshore Accords and their enabling statutes, provincial social laws apply offshore Newfoundland and Nova Scotia, respectively. Again, given industry's experience elsewhere in the country, this will not necessarily cause major difficulties. A larger problem is that, as of April 1989, there is no statute which extends the applications of laws generally to the offshore. This matter is explored more fully in the section entitled "Legislative Authority".

## Relationships Between the Offshore Petroleum Boards, the Federal and Provincial Governments, and the Federal and Provincial Agencies

The offshore regimes have resulted (or, in the case of Nova Scotia, will result) in the creation of two new bodies, the Offshore Petroleum Boards (the CNOPB and the CNSOPB). It will be necessary for these institutions to find their niche in the federal-provincial web of relationships that already exist. The fact that the Accords have resulted in a transfer of some authority from the ministers to these Boards will also require some adjustment by the various actors involved in offshore petroleum management. And, as was explained earlier, the nature of the Accord system dictates the need for constant consultation at several levels: between the federal government and each of the Boards; between the federal government and each of the provincial governments; between each provincial government and its Board; and, ideally (but not necessarily), between the two Boards and between the two provincial governments. If intercommunication at all these levels works smoothly and effectively, some of the inherent problems caused by the cumbersome and complex nature of the system may be avoided. If there are breakdowns in these important lines of communication, irritations and inefficiencies may be inevitable. All this implies an important role for COGLA as facilitator, communicator and coordinator, for it is in a good position to supply the glue to hold the system together.

The Accords foresaw the need to develop new relationships between the Offshore Petroleum Boards and various government departments. Clause 63 of the *Atlantic Accord* and 40.01 of the *Nova Scotia Accord* provided that the Boards would conclude Memoranda of Understanding [hereinafter MOU] with government departments and agencies having continuing responsibilities in the offshore, "with a view to ensuring effective coordination and minimum duplication." The enabling legislation contains more explicit provisions to the same effect.[55] In Newfoundland, five such MOU are under negotiation (dealing with the fisheries, the environment, the Coast Guard, industrial and employment benefits, and occupational health and safety). In each case, the signatories will include the CNOPB, the two energy departments, the departments of intergovernmental affairs/relations, and other relevant departments. The CNOPB has taken the lead in initiating the preparation of the MOU. The objective of all these MOU is to set out how the Board can provide a single window for the industry.

While relationships between the federal and Newfoundland governments and the CNOPB seem to have worked quite smoothly, communication between Nova Scotia and Newfoundland is less apparent. Part of this may reflect earlier rivalry between the two provinces and historically different styles of approaching the federal government. In addition, the absence of a Board in Nova Scotia has made it impossible so far to develop the relationships that will be desirable. Although there are already examples of good tripartite relationships (for example, in relation

55. *See*, *Atlantic Accord Act*, *supra*, note 23, s.46; *Nova Scotia Accord Act*, *supra*, note 41, s.46.

to rights issuance), ongoing efforts will be required to build lines of communication between the two Boards, thereby minimizing deviations in practices and policies.

The relationship between the ministers and the Offshore Petroleum Boards in relation to fundamental decisions may prove important. As mentioned earlier, the CNOPB has already made a number of fundamental decisions without interference from either minister. The 1987-88 controversy in Nova Scotia concerning Texaco's application to drill on Georges Bank, however, demonstrates the sort of complexities that could arise in future if the views of the two governments do not coincide on a fundamental decision.

Industry interests in the Georges Bank area date back to the mid-1960s. While seismic and similar activities have taken place, there has been no drilling: a long-standing boundary dispute between Canada and the United States was resolved by the International Court of Justice in October 1984, and, in the interim, the Canadian government had imposed a moratorium on drilling activities in the disputed area. Following resolution of the boundary dispute, Texaco expressed interest in mounting a drilling program pursuant to rights obtained over two decades ago. In June 1986, COGLA released a preliminary review of the environmental implications of a drilling program on Georges Bank, concluding that, subject to specific recommendations, a limited exploratory drilling program could proceed.[56]

Texaco undertook a process of public consultation concerning its proposed program and commissioned additional environmental studies. In a press release of November 24, 1987, however, the Premier of Nova Scotia announced that, due to concerns expressed by the fishing industry about the proposed drilling, the province had decided to suspend administrative processes relating to Georges Bank pending a public hearing by the CNSOPB (the establishment of which was then expected to occur shortly, but

---

56. Canada - Nova Scotia Offshore Oil and Gas Board, *An Environmental Evaluation of the Proposed Exploratory Drilling on Georges Bank* (1986).

which as of April 1989, has not yet been established). He indicated that if, following such public hearings, the CNSOPB decided to grant such exploration licences, the province would veto the decision and request the federal minister to do likewise. Noting his authority to veto a development plan, he also stated that he would exercise that right to prevent any drilling or development on Georges Bank.

In January 1988 the Nova Scotia energy minister approached the federal energy minister with a request that a permanent moratorium be placed on oil and gas activities on Georges Bank. Three months later, the federal energy minister proposed that a moratorium on drilling on Georges Bank be imposed at least until the year 2000 and that a thorough public review of the issue be conducted.[57]

For the present, the Georges Bank issue has been defused by federal-provincial negotiation and agreement. It nevertheless reminds us that the two levels of government may not always have the same view on contentious issues. The stance adopted by the Nova Scotia Premier in the early stages of this controversy calls

---

57. The Honourable Marcel Masse, M.P. for Frontenac and Minister of Energy, Mines and Resources Canada, Address (Halifax, N.S., 18 April 1988). During the time of this controversy, the federal implementing legislation was being considered in a Parliamentary committee. The arrangement for the moratorium was subsequently added to the Bill for the *Nova Scotia Accord Act*, *supra*, note 41, s.141. The province's *Nova Scotia Implementation Act* had already been passed, but was unproclaimed. It was amended by Bill No. 110, *An Act to Protect Georges Bank*, which was given first reading on May 13, 1988, and received royal assent on May 25, 1988 (now S.N.S. 1988, c.56). It has been observed that the decision not to allow drilling on Georges Bank could have been handled under the already-agreed procedures for offshore management, without the necessity of legislating a moratorium. *See* Rowland J. Harrison, "The Changing Offshore Petroleum Management Regime on the East Coast: Its Responsibilities and Functions" (Luncheon Address to the 1988 CORE Conference, Halifax, 6 October 1988). Moreover, it is far from clear that, had no compromise been achieved between the two levels of government, the Nova Scotia Premier would have been legally entitled to take the action he outlined, i.e., veto a Board decision to issue exploration licences. A full legal analysis of this issue would have to consider the status of the earlier Texaco rights under the current legislation.

into question whether the Offshore Petroleum Boards can be truly independent in the face of political differences and illustrates the close control retained over the offshore by politicians.

Legislative Authority

Prior to the Accords, the existence of two sets of governing legislation administered by two levels of government sometimes created an almost untenable situation for offshore operators.[58] Major strides forward have been made through the Accords, with the incorporation by reference of specific provincial laws and the development of a procedure under which mutually-agreeable regulations can be promulgated. Nevertheless, there remain some areas of concern.

The largest problem is the continuing absence of an overall "offshore laws application" statute. The need for such a law has been long recognized since, under Canadian law, laws do not apply beyond the low water-mark unless they are expressly stated to do so.[59] The Royal Commission on the *Ocean Ranger* Marine Disaster, in its second report, recommended the passage of such a law;[60] in response, the federal government introduced Bill C-104.[61] The *Harrison Task Force Report* expressed doubt about whether the proposed Bill was the best mechanism for dealing with this matter, and, in addition, some of the provinces objected to the form

---

58. For an example of one such situation, *see* Constance D. Hunt, "Analysis of Constitutional Issues for Energy Projects in Canada" *in International Energy Law* (New York: Matthew Bender, 1984).
59. For a discussion of the laws that had been applied to the offshore as of 1983, *see* A.L.C. de Mestral, "The Law Applicable to the Canadian East-Coast Offshore" (1983) 21 Alta. L. Rev. 63. *See also* Lawrence L. Herman, "The Need for a *Canadian Submerged Lands Act*: Some Further Thoughts on Canada's Offshore Mineral Rights Problems" (1980) 58 Can. B. Rev. 518.
60. Canada, Royal Commission on the *Ocean Ranger* Marine Disaster, *Report Two: Safety Offshore Eastern Canada* (Ottawa: 1985) Recommendation #78.
61. Bill C-104, An Act to apply federal laws and provincial laws to offshore areas and to amend certain Acts in consequence thereof, 1st session, 33d Parl., 33-34-35 Eliz. II, 1984-85-86. The Bill died on the order paper.

and content of the legislation.[62] The Bill died on the order paper and no further proposals have been introduced. The resulting gaps in the statutory coverage for the offshore remain a concern for industry.

The interrelationship between certain provincial and federal laws has been clarified under the Accords, although some potential confusion about overlap remains. If an offshore structure or installation is intended to be fixed, s.152(4)(b)(i) of the *Atlantic Accord Act* makes the Newfoundland *Labour Relations Act, 1977*, applicable rather than Part V of the *Canada Labour Code*[63] (the certification provisions), "during such time as the marine installation is within the offshore area" in connection with the exploration, drilling, production, conservation, or processing of petroleum. If a gravity-based system is being constructed on land, the workers will, of course, be subject to provincial labour laws during the construction phase. As a result, provincial certification laws will apply to the construction, towing, and operation phases of a fixed offshore structure. Federal certification laws will continue to apply to mobile units, preventing the necessity of union recertification when the unit crosses a provincial boundary. There remains some doubt, however, about what labour laws will apply to other workers, such as those in supply boats: the definition of "marine installation or structure" in s.152(1) of the *Atlantic Accord Act* includes ships, while s.152(2) makes Newfoundland social legislation (including *The Labour Standards Act*) applicable to marine installations or structures in the offshore in connection with petroleum activities.

Section 152(4)(a) of the *Atlantic Accord Act* makes Parts III

---

62. Potentially, the Bill affected every province except Saskatchewan and Alberta. Many of the coastal provinces take the view that their laws already apply and do not require a federal statute to have this effect.
63. *Canada Labour Code*, R.S.C. 1985, c.L-2. The counterpart provision to s.152(4)(b)(i) in the *Nova Scotia Accord Act* is s.157(4)(b)(i). I am grateful to Tamara Parschin-Rybkin of COGLA for helping me to understand the intricacies of the provisions dealt with in this and the two following paragraphs.

and IV (occupational health and safety) of the *Canada Labour Code* inapplicable to marine installations or structures in the offshore area in connection with petroleum activities.[64] Again, if an installation or structure is constructed on land, provincial laws will apply during the construction phase. Once moved to sea, however, regulations made pursuant to the joint process will apply.[65] Some confusion arises from the interrelationship of s.152(2) and (3) of the *Atlantic Accord Act*.[66] The first provides for the application to installations of Newfoundland social laws and regulations. The second states:

> Notwithstanding subsection (2), any provision of any Act or regulation referred to in that subsection made in relation to a matter respecting which a regulation under paragraph 149(1)(d), (m), (o) or (p) or any other provision of this Act respecting occupational health or safety *may be made* does not apply on marine installations or structures referred to in subsection (2) during such time as those structures are within the offshore area in connection with a purpose referred to in that subsection. [emphasis added]

This provision causes problems because of the general reaction: "Why was the province given jurisdiction over social legislation with one hand only to have it taken away with the other?" The reasoning behind this provision is that many offshore-specific regulations made through the joint process will be better suited to the offshore situation than the province's general social laws. While this is a sensible approach, the wording may be problematic so long as joint regulations are not in place. Note that the section does not refer to regulations that "have been made" under the *Atlantic Accord Act*, but rather to regulations that "may be made". The result seems to be that, if regulations have not been passed, there will be a gap in these laws in relation to offshore

64. The counterpart provision in the *Nova Scotia Accord Act* is s.157(4)(a).
65. The joint process for making occupational health and safety laws is found in s.152(3) of the *Atlantic Accord Act* and s.157(3) of the *Nova Scotia Accord Act*.
66. The same analysis applies to s.157(2) and (3) of the *Nova Scotia Accord Act*.

installations.[67] From this analysis concerning the social laws, it may be concluded that, while the Accord legislation is a major improvement to the pre-Accord situation, the rather convoluted way in which social legislation has been made applicable to the offshore is likely to cause some continuing confusion.

Regulations that pertain to Mobile Offshore Drilling Units [hereinafter MODUS] may also pose future problems. The Accord legislation contemplates the passage of jointly-agreed regulations on this topic.[68] Similar regulations are authorized by 1987 amendments to the *Canada Shipping Act*.[69] While regulations pursuant to the latter would only apply to rigs if the federal energy and transport ministers agreed, this possibility will need to be monitored to avoid duplication.

In relation to MODUS generally, it seems that the Accord regime has not solved issues concerning the role of the Canadian Coast Guard. It is said that inspectors appointed by the Coast Guard and those appointed for the purpose of the Accord legislation do not always take the same view. While this administrative overlap is not new, it is disappointing that it remains unresolved under the Accords.

## Review of Decisions

Section 30 of the *Atlantic Accord Act* and s.31 of the *Nova Scotia Accord Act* make the exercise of a power or the performance of a duty by the Offshore Petroleum Boards final and not subject to

---

67. One might argue that the same situation pertains to many aspects of the offshore: since very few regulations have been passed yet, much of the activity is regulated pursuant to draft regulations. The *Harrison Task Force Report* commented on the fact that most of the regulations remain in draft form. See *Harrison Task Force Report*, *supra*, note 52, at 11.
68. The *Atlantic Accord Act*, *supra*, note 23, s.149(1)(p); the *Nova Scotia Accord Act*, *supra*, note 41, s.153(1)(p).
69. *Canada Shipping Act*, S.C. 1987, c.7, s.46. This adds s.400.1 to the *Canada Shipping Act*, authorizing the Governor in Council to make regulations to implement, *inter alia*, the Code for the Construction and Equipment of Mobile Offshore Drilling Units (1979) and to impose stricter standards.

review or approval of either government or minister, "subject to this Act". Many of the decision-making powers of the Boards are subject to ministerial involvement. This section is concerned with two other types of review: that of the Oil and Gas Committee and the courts. Reference hereinafter is to the *Atlantic Accord Act*; unless indicated otherwise, it can be assumed that there are identical provisions in the *Nova Scotia Accord Act*.

The appointment of an Oil and Gas Committee is contemplated by Part III of the Act which deals with production and conservation matters. The notion of such a Committee originates with the OGPCA, where it was designed to deal with matters such as pooling and unitization that were thought likely to arise on lands in the Northwest Territories. Though a Committee was constituted under the OGPCA, it never met and was eventually disbanded.[70] A study concerning oil and gas conservation on federal lands identified a number of problems with the concept of the Committee and the legislative provisions in OGCPA that provide for it.[71]

Despite the fact that the Committee contemplated by the OGPCA had never been utilized, when the CPRA was passed, the Committee was assigned new powers relating to the exploration stage of petroleum activities.[72] These provisions, along with the OGPCA provisions, were duplicated in the legislation enabling the Accords with one major change: under the CPRA and the OGPCA, the Committee is to oversee ministerial decisions,

---

70. For a discussion of the origins and early history of the Committee, *see* "Canada Oil and Gas Committee" *in* Constance D. Hunt & Alastair R. Lucas, *Canada Energy Law Service* (Don Mills: Richard de Boo, 1981- ). A Committee is about to be constituted to deal with an issue that has arisen in the Arctic.
71. *See* Owen L. Anderson, *Oil and Gas Conservation on Canada Lands* (Calgary: Canadian Institute of Resources Law, University of Calgary, 1985).
72. The CPRA, *supra*, note 26, s.106, assigned the Committee the power to advise the minister as to his decisions in relation to significant discovery declarations (s.28), drilling orders (s.33), commercial discovery declarations (s.35), interest reduction orders (s.36), and cancellation of rights (s.105).

whereas under the Accord legislation the Committee must interact with the Boards. This marriage of an old concept with new administrative arrangements has left the Committee with a confused role.

The establishment and organization of the Oil and Gas Committee are dealt with in ss.141 to 148 of the *Atlantic Accord Act.* The establishment of a Committee by the CNOPB is not mandatory, even though the industry has been granted the *right* to have certain of the CNOPB's decisions under Part II of the Act reviewed by a Committee (see Table A - Part IV and discussion below). This is but one of the anomalies that surround the Committee. The Committee is to be composed of up to five members, of which not more than three are to be members of the public service of Canada or a province and at least two of whom are to have specialized, expert, or technical knowledge of petroleum. Pursuant to s.147, the CNOPB may refer to the Committee for advice "any question, matter or thing arising under this Part or relating to the conservation, production, storage, processing or transportation of petroleum." Section 145(2) gives the Committee the powers, rights, and privileges of a superior court of record, while s.148(1) makes its orders enforceable in like manner as an order of the Supreme Court of Newfoundland. By virtue of s.145(3), "[t]he finding or determination of the Committee on any question of fact within its jurisdiction is binding and conclusive."

It has been seen that an affected party is entitled to an Oil and Gas Committee hearing to review a CNOPB decision in relation to a number of Part II matters of potentially great significance to the industry (including interest reduction orders, drilling orders, cancellation of rights, and declarations of significant and commercial discoveries). The Committee's role *vis-a-vis* the CNOPB in this s.124 procedure is advisory. Section 124(10) also contemplates judicial review of these matters:

> Any order, decision or action in respect of which a hearing is held under this section is subject to review and to be set aside by the Trial Division of the Supreme Court of Newfoundland.

Note that no grounds for judicial review have been specified.

In addition to its advisory powers under Part II and its general advisory role pursuant to s.147, the Oil and Gas Committee has appellate powers over certain decisions of the Chief Conservation Officer, such as waste orders.[73] Moreover, ss.168, 173, and 176 give the Committee the authority to make orders respecting pooling and unitization.

Division III of Part III of the *Atlantic Accord Act* (Appeals and Administration) further confuses the already complex role of the Oil and Gas Committee. Section 184(1) states: "Except as otherwise provided in this Division, every decision or order of the Committee is final and conclusive." This is contradictory because, as explained above, the Committee's Part II functions are made judicially reviewable by s.124(10). Section 185 permits the Committee on its own motion or at the request of the CNOPB to state a case to the Trial Division of the Supreme Court of Newfoundland "on any question that in the opinion of the Committee is a question of law or of the jurisdiction of the Committee". A decision or order of the Committee can also be appealed to the Court pursuant to s.187 on a question of law. Finally, s.186 gives the CNOPB the power to rescind or vary any order or decision of the Committee made under Part III. Oddly, this power seems to exist even when the Committee's decision has been reviewed by the Court on a matter of law, so long as the CNOPB has not already dealt with that order.[74] This suggests that the CNOPB could, in effect, overrule a decision of the Court.

This brief review suggests that there are serious problems with the legislation in regard to the interaction of the CNOPB, the Oil and Gas Committee, and the courts. This problem has been broadly recognized. Indeed, Newfoundland has already put

---

73. *Atlantic Accord Act*, *supra*, note 23, s.157.

74. *Atlantic Accord Act*, *supra*, note 23, s.187(4) provides: "Any order made by the Committee pursuant to subsection (3), unless that order has already been dealt with by the Board pursuant to section 186, shall be subject to that section". Subsection (3) requires the Committee to make an order to comply with an opinion of the Court.

forward proposals for amendments and the role of the Committee is being examined in the review of the OGPCA. The situation is explained by the fact that there was a deadline to be met in passing the Accord legislation, which was complicated by the drafting of the CPRA. As it proved impossible to cope with the OGPCA's deficiencies in the necessary time-frame, it was decided to leave those matters for later resolution.

## 3. AUSTRALIA

### Background

Australia has had a great deal of experience in offshore petroleum production.[75] While some exploration began prior to the turn of the century, efforts were sporadic and commercial deposits of oil and natural gas were not found on land until the 1960s. A major offshore success occurred in the mid-1960s with discoveries of oil by the Broken Hill Proprietary Company Ltd. [hereinafter BHP] and Esso Exploration and Production Australia Inc. in the Gippsland Basin off the Victorian coast. These fields are now in decline, but still continue to produce three-quarters of Australia's indigenous oil.[76] There is also a small amount of offshore hydrocarbon production from the Carnarvon Basin in state waters off Western Australia and from the federal territory of Ashmore and Cartier Islands off northwest Australia. Natural gas discoveries in the continental shelf of northwest Australia and limited oil discoveries in the same area have resulted in growing exploratory interest there.

The Commonwealth of Australia consists of six states and several federal territories, one of which (the Northern Territory) now enjoys a considerable degree of self-government. There are

---

75. For an overview of Australia's petroleum history up until 1971, *see* Australia, Senate Select Committee, *Report from the Senate Select Committee on Offshore Petroleum Resources* (Canberra: Commonwealth Government Printing Office, 1971) [hereinafter *Senate Select Committee Report*] c.3.
76. For an overview of current offshore petroleum exploration and production, *see* Robert J. Foster, "Australian Crude Oil in a Free Market" *in* International Bar Association, Section on Energy and Natural Resources Law, *Energy Law '88: Pre-Seminar Papers* (Sydney: 1988) at 13. Details on activities in Western Australia may be found in Western Australia, Dept. of Mines, *Petroleum in Western Australia* (1988) at 19.

important differences of history and geography between Canada and Australia that have influenced the development of their respective offshore petroleum regimes. For example, the origins of the Commonwealth of Australia are six colonies, all of which simultaneously became part of the federation in 1901; in contrast, not all the Canadian provinces joined Confederation at the same time or under the same circumstances. All six states and the Northern Territory have lengthy coastlines, while two of the Canadian provinces are totally landlocked.

Some of the constitutional differences between the two countries should also be noted. In Canada, there is a division of exclusive legislative authority between the two levels of government, with residual power falling to the federal government pursuant to the "peace, order and good government" clause. In Australia, legislative powers allocated to the two levels of government are, for the most part, concurrent rather than exclusive. Section 51 of the *Commonwealth Constitution Act*[77] lists matters with respect to which the federal government has power to make laws "for the peace, order, and good government of the Commonwealth", including interstate and international trade and commerce, taxation, defence, fisheries beyond territorial limits, and external affairs. Section 106 continues the Constitution of each state as at the establishment of the Commonwealth. In general, state Parliaments have potentially wider legislative authority than the Commonwealth Parliament, as they retain the residue of powers not specifically assigned to the Commonwealth. However, s.109 provides: "When a law of a State is inconsistent with a law of the Commonwealth, the latter shall prevail, and the former shall, to the extent of the inconsistency, be invalid." As a result, where a legislative subject falls within Commonwealth jurisdiction and the Commonwealth exercises its authority, the legislative power of the state is circumscribed accordingly.

---

77. *Commonwealth of Australia Constitution Act*, 63-64 Vict., c.12 [U.K.] [hereinafter *Commonwealth Constitution Act*].

The 1967 Agreement

Discussions relating to intergovernmental arrangements for offshore petroleum commenced in the early 1960s.[78] In the meantime, offshore exploration was authorized by state and territory governments pursuant to onshore petroleum legislation. Although there was no clear legal authority concerning jurisdiction over offshore resources, by 1964 "the States had arrived at the firm view that they should continue to control oil and mineral rights in their off-shore adjacent areas", but had decided to avoid conflict with the Commonwealth over continental shelf rights.[79]

Following three years of intergovernmental discussions, the *1967 Agreement*[80] was reached concerning a cooperative scheme for the management of offshore petroleum. The Agreement was signed by the Commonwealth and state governments on October 16th when, at the national level, a Liberal-Country Coalition controlled the House of Representatives, but not the elected Senate. The preamble to the Agreement notes that offshore petroleum activities would be encouraged by the existence of uniform legislative measures applying to the continental shelf and areas beneath territorial waters; that the governments had decided, in the national interest, to cooperate to ensure the effectiveness of authorities over petroleum resources, without raising questions concerning and without derogating from their respective constitutional powers; and that the governments had agreed to submit the necessary legislation to their respective Parliaments and to cooperate in the administration of such legislation.

The key provisions of the *1967 Agreement*'s twenty-six clauses may be summarized as follows:

---

78. For a more complete background on these discussions, see *Senate Select Committee Report*, *supra*, note 75, at c.IV.
79. *Id.*, at 75.
80. *Agreement relating to the Exploration for, and the Exploitation of, the Petroleum Resources, and certain other Resources, of the Continental Shelf of Australia and of certain Territories of the Commonwealth and of certain other Submerged Lands*, reprinted as Appendix A *in Senate Select Committee Report*, *supra*, note 75, at 714 [hereinafter *1967 Agreement*].

- While the Agreement was not intended "to create legal relationships justiciable in a Court of Law", it would be varied, revoked, or determined only by the agreement of all parties (clauses 26 and 25);
- Each party agreed to submit to its Parliament the so-called "Common Mining Code" which was appended as a Schedule to the *1967 Agreement.* This would result in virtually identical Commonwealth and state legislation to govern offshore petroleum activities: the "Commonwealth Mining Code" would apply through the entire offshore while the "Common Mining Code" of each state would apply to its own "adjacent area". Alteration of these "mirror" statutes would take place only by agreement of the governments, and regulations pursuant to the statutes would be made, amended, or repealed only by agreement (clauses 3 to 7);
- The offshore area adjacent to each state would be administered by a Designated Authority who, typically, would be a state minister of mines, energy, or resources (clause 9);
- Unless advised by the Commonwealth minister that it was unnecessary to do so, the state government would consult the Commonwealth government prior to the issue, renewal, or variation of petroleum interests and prior to granting approval for the transfer or creation of new interests. The decision of the Commonwealth minister would be made taking into account only listed matters of Commonwealth authority (such as taxation and international trade and commerce), and the state governments undertook to give effect to a Commonwealth decision thus made (clause 11);
- A Designated Authority ordinarily would not give a direction inconsistent with regulations made under the "Common Mining Code" nor exempt an interest holder from compliance except after consultation with the Commonwealth minister (clause 18);
- A maximum ten per cent royalty would be split on a 40/60 per cent basis between the Commonwealth and the adjacent

state, with the state being entitled to any "over-ride" royalty above ten per cent and to any other moneys payable under the "Common Mining Code" (clauses 19 and 22);

- There would be ongoing consultation between the governments concerning the *1967 Agreement* (clause 23).

Described by a Canadian critic as a "unique and extraordinary constitutional document",[81] the *1967 Agreement* was implemented at the Commonwealth level primarily by the *Petroleum (Submerged Lands) Act 1967*[82] [hereinafter PSLA]. The PSLA did not sail smoothly through the required legislative waters. The necessary Bills passed the House of Representatives and were transmitted to the Senate, where the Opposition proposed an amendment referring the PSLA and its associated Bills to a Senate Select Committee. As a compromise, it was agreed that the Senate would pass the Bills and then establish a Senate Select Committee to inquire into matters concerning offshore petroleum resources. While this Committee was deliberating, the state governments passed the enabling legislation required by the *1967 Agreement*[83] and offshore petroleum activities proceeded according to the cooperative scheme.

## The Report of the Senate Select Committee on Offshore Petroleum Resources

The Senate Select Committee was charged to inquire into and

---

81. Andrew Thompson, "Australia's Offshore Petroleum Common Code" (1968) 3(2) U.B.C. L. Rev. 1 at 4.
82. The accompanying Commonwealth legislation was: *Petroleum (Submerged Lands) (Royalty) Act 1967* (Cth) (No.119) [hereinafter *Royalty Act*], *Petroleum (Submerged Lands) (Exploration Permit Fees) Act 1967* (Cth) (No.120), *Petroleum (Submerged Lands) (Production License Fees) Act 1967* (Cth) (No.121), *Petroleum (Submerged Lands) (Pipeline License Fees) Act 1967* (Cth) (No.122), *Petroleum (Submerged Lands) (Registration Fees) Act 1967* (Cth) (No.123), *Petroleum (Ashmore and Cartier Islands) Act 1967* (Cth) (No.124).
83. The main Victorian and Western Australian statutes were *Petroleum (Submerged Lands) Act 1967* (Vict.) (No.7591) and *Petroleum (Submerged Lands) Act 1967* (W.A.) (No.41).

report upon seven specific matters. Its deliberations spanned three and a half years, embracing 183 sessions and evidence from 84 witnesses. A decade and a half after its release, this voluminous *Report* is still a comprehensive source of information concerning issues related to the management of offshore petroleum resources.

Some of the Senate Select Committee's observations will be referred to in detail later, but a few require specific mention at this point. While recognizing certain advantages in the offshore arrangement, the Committee concluded that it was not in the national interest to leave "unresolved and uncertain the extent of state and Commonwealth authority in the territorial sea-bed and the Continental Shelf".[84] It perceived some challenge to the exercise of parliamentary functions in the notion of "uniform legislation drafted by the executive arms of seven Australian Governments being presented to the Parliaments as a *fait accompli* requiring formal legislative approval".[85] It saw a need to reduce the scope of discretion vested in the Designated Authorities and expressed concern about the fact that the accountability or responsibility of ministers with decision-making authority under the legislation was incomplete.[86] It also noted that there had been difficulties in procuring expeditious and coordinated amendments to the mirror Commonwealth-state legislation.[87]

It is unclear exactly what role was played by the *Senate Select Committee Report* in the events which followed. Given the thoughtfulness and depth of the *Report* and the extent to which it is still quoted, it must have had considerable influence. Still, between 1967 (when the Agreement was reached) and 1971 (when the *Report* was released), there had been important new developments. For example, the Supreme Court of Canada had handed down its decision in the *B.C. Offshore Reference*,[88] lending support to those who believed in the strength of the

---

84. *Senate Select Committee Report*, *supra*, note 75, at 7.
85. *Id.*
86. *Id.*, at 9.
87. *Id.*, at 10.
88. *B.C. Offshore Reference*, *supra*, note 11.

Commonwealth's legal position relating to offshore jurisdiction. In 1970 when the Liberal, Mr. Gorton, became Prime Minister, he introduced the Territorial Sea and Continental Shelf Bill, affirming federal jurisdiction over both the territorial sea and the continental shelf. Although the Bill did not pass (in part due to disagreement within the Liberal Party), it was supported by the opposition Labour Party,[89] demonstrating that the offshore issue remained contentious.

The 1973 Seas and Submerged Lands Act
and the 1975 Seas and Submerged Lands Case

Shortly after its election to national office in 1973, the Labour government introduced its own *Seas and Submerged Lands Act 1973*,[90] which, like the ill-fated Territorial Sea and Continental Shelf Bill, declared federal sovereignty over the continental shelf and the territorial sea. Having unsuccessfully attempted to have the matter of offshore jurisdiction referred to the Privy Council prior to the introduction of this legislation,[91] the states collectively attacked its validity before the High Court.

The High Court decision in the *Seas and Submerged Lands Case*[92] was handed down in 1975. All seven judges upheld the validity of the *Seas and Submerged Lands Act 1973 vis-a-vis* the continental shelf by virtue of the Commonwealth's legislative authority in relation to external affairs. Five judges upheld the provisions relating to matters other than the continental shelf on the same ground, while two judges would have found invalid the provisions relating to the territorial sea.

In the period between the passage of the *Seas and Submerged*

---

89. Richard Cullen, *Australian Federalism Offshore*, 2d ed. (Melbourne: Intergovernmental Relations in Victoria Program, Law School, University of Melbourne, 1988).
90. *Seas and Submerged Lands Act 1973* (Cth) (No.161).
91. Cullen, *supra*, note 89, at 40.
92. *New South Wales* v. *The Commonwealth* (1975), 135 C.L.R. 337 (Aust. H.C.) [hereinafter *Seas and Submerged Lands Case*].

*Lands Act 1973* and the High Court decision, the cooperative arrangements that had characterized the *1967 Agreement* gradually fell apart.[93] For example, the Commonwealth government announced that it would not approve new farm-ins, a power it was able to exercise by virtue of the provision in the *1967 Agreement* requiring the Designated Authority to consult the Commonwealth government before approving transfers. The Commonwealth position on this matter reflected its desire to increase Australian ownership in offshore projects by itself becoming a partner in developments. Problems also arose in regard to the flow of information between the two levels of government. Furthermore, the Commonwealth minister suggested that no permits should be renewed and no relinquished areas re-issued until after the High Court's decision. The state governments collectively defied this request, and, as happened later in Canada, industry often found itself caught between two warring governments.

With the High Court decision, it was apparent that new arrangements were required and a lengthy round of intergovernmental negotiations ensued. The results led eventually to Australia's *1979 Offshore Constitutional Settlement* regarding offshore petroleum.

## The 1979 Offshore Constitutional Settlement

Terms of the 1979 Offshore Constitutional Settlement

In contrast to the *1967 Agreement*, no formal document evidencing the *1979 Offshore Constitutional Settlement* [hereinafter *1979 OCS*] has ever been released. Instead, the terms of the *1979 OCS* must be gleaned from a package of information circulated by the Commonwealth Attorney General's Department.[94]

---

93. The matters mentioned here are discussed in more detail in Garth Stevenson, *Mineral Resources and Australian Federalism*, Research Monograph No.17 (Canberra: Centre for Research on Federal Financial Relations, Australian National University, 1976) at 41 *ff*.
94. The information kit, entitled "Offshore Australia", includes maps and two booklets: Australia, Dept. of the Attorney General, *Offshore*

The form in which the OCS was expressed has had certain implications which will be explored in the section entitled "The Canadian and Australian Offshore Petroleum Regimes Compared".

The following terms of the *1979 OCS* are of particular interest to this study:

- The Commonwealth would pass legislation giving each state[95] the same powers with respect to its adjacent territorial sea as it would have if the waters were within the state.
- The Commonwealth would pass legislation vesting in each state proprietary rights and title in respect of the seabed of its adjacent territorial sea. This was intended to support the extension of state power to the territorial sea and "provide an assurance to the States that the arrangements relating to the territorial sea will have permanency and stability".[96] Commonwealth powers to use the seabed for such matters as defence and navigational aids would be reserved.
- The state territorial sea would be limited to a breadth of three miles. Baselines would be drawn in consultation with the states and promulgated under the *Seas and Submerged Lands Act 1973*.
- Inside the territorial sea, state petroleum legislation would govern, with the "Common Mining Code" retained "as far as practicable".[97]
- Outside the three-mile territorial sea, petroleum arrangements

---

*Constitutional Settlement: A Milestone in Co-operative Federalism* (Canberra: Australian Government Publishing Service, 1980) (reproduced in part in Michael Crommelin, "Offshore Mining and Petroleum Constitutional Issues" (1981) 3 A.M.P.L.J. 191) and Australia, Dept. of the Attorney General, *Offshore Constitutional Settlement: Selected statements and documents*. The "Agreed Arrangements" contained in the former are reproduced as Appendix A in Crommelin, *id.*

95. Under the *1979 OCS*, the Northern Territory is treated very much like the states. Since this study is concerned primarily with the states of Victoria and Western Australia, there is little reference hereinafter to the Northern Territory or to other federal territories. It can generally be assumed that the arrangements for the states also apply to the Northern Territory.
96. *Offshore Constitutional Settlement, supra*, note 94, at 6.
97. *Id.*, at 8.

would be regulated by an amended PSLA, with day-to-day administration in the hands of a Designated Authority (the state minister). Certain matters, such as the opening of areas for permits, the grant and renewal of exploration permits and production licences, the approval of instruments creating interests in permits or licences, and the determination of certain licence or permit conditions would be decided by a Joint Authority consisting of the relevant state and the Commonwealth ministers, with the Commonwealth minister having the final say in the event of disagreement. Western Australia would have a slightly broader role in decision -making than the other states.

- The existing arrangements for the sharing of royalties between the states and the Commonwealth would be retained.[98]
- Other arrangements were made with regard to fisheries, historic shipwrecks, marine parks, criminal law, and shipping and navigation. In relation to the latter, the Commonwealth would be responsible for the navigation and marine aspects of offshore industry mobile units (mainly drilling vessels), although *Navigation Act* requirements might be displaced by "directions" or "conditions of instruments" issued under the PSLA. The Commonwealth would also be responsible for offshore industry vessels (mainly supply craft) other than those confined to one state and its adjacent area. Again, PSLA

---

98. The booklet itself does not make it clear that the 40/60 per cent sharing of royalties would continue to apply within the territorial sea, since the reference to a continuation of the arrangements is contained in a section of the booklet that relates to the area beyond the three-mile territorial sea. Moreover, the booklet refers only to a continuation of the royalty-sharing arrangements and does not mention the provision in the *1967 Agreement* pursuant to which the states were entitled to receive all moneys other than royalty. Nevertheless, it seems to be assumed that the royalty-sharing arrangement continues to apply to the territorial sea and that the states are entitled to other moneys. See, for example, Australia, Dept. of Resources and Energy, *'Energy 2000' —A National Energy Policy Review: Petroleum* (1986) at 11 and *Petroleum (Submerged Lands) Act 1982* (Vict.) (No.9772), s.129, which provides for payments to the Commonwealth by the state in relation to petroleum production within the territorial sea.

"conditions" and "directions" could replace *Navigation Act* requirements. Owners of offshore industry vessels would be required to make a declaration concerning the intended use of the vehicle for the purpose of determining whether its use would be confined to one state.

- As to ship-sourced marine pollution, the existing arrangements would be continued whereby state legislation would apply within the territorial sea and Commonwealth legislation beyond, with Commonwealth authorities retaining the power to intervene in certain cases of threatened or actual oil pollution.

Legislative Implementation

The *1979 OCS* was implemented through Commonwealth and state statutes. As outlined above, the OCS goes beyond petroleum matters, but here the emphasis will be on the petroleum-related legislation.

Two Commonwealth statutes form the basis of the *1979 OCS*. First, the *State Title Act*[99] vests title in each state to its "coastal waters", giving each state the same right and title to the property in the seabed beneath the coastal waters and the space above that seabed as if the seabed were beneath waters of the sea within state limits. The term "coastal waters" is defined to include a state's internal waters and a three-mile territorial sea.[100] Commonwealth rights are protected by s.4(2) in relation to such matters as communication, defence, navigation safety, and the construction of pipelines for the transport of petroleum recovered from areas beyond state coastal waters. Section 4(5) gives the states the power to dispose of or otherwise deal with the vested area in

---

99. *Coastal Waters (State Title) Act 1980* (Cth) (No.77) [hereinafter *State Title Act*].

100. In s.3(1) of the *State Title Act*, *id.*, "coastal waters of the State" is defined by reference to the same expression as in the *Coastal Waters (State Powers) Act 1980* (Cth) (No.75) [hereinafter *State Powers Act*]. The detailed definition is found in s.3(1) of the *State Powers Act*.

accordance with state law.

Second, the *Coastal Waters (State Powers) Act 1980*[101] [hereinafter *State Powers Act*] provides that the legislative powers of each state under its Constitution extend to the coastal waters of that state as though those areas were within a state's territorial limits. In relation to certain matters such as ports, harbours and shipping facilities, state legislative powers extend beyond the coastal waters.

Extensive amendments were made to the Commonwealth PSLA[102] to ensure that its provisions would not interfere with the operation of the *State Powers Act* and to implement other elements of the *1979 OCS* (such as providing for the role of the Joint Authority). While the PSLA had applied originally throughout the offshore area, the amendments made it applicable only in the "Commonwealth adjacent area" beyond the territorial sea.[103] New or amended state-PSLAs were passed to govern activities within the state coastal waters, to make transitional arrangements for existing permits, and to provide for a "Common Mining Code".

To summarize, the *1979 OCS* was implemented by the Commonwealth PSLA that granted to each state title and legislative authority over its "coastal waters" (internal waters and a three-mile territorial sea). The Commonwealth PSLA was amended to make it applicable only in the "Commonwealth adjacent areas" (the areas offshore each state and beyond state coastal waters). Legislative uniformity over petroleum activities in both the state and Commonwealth areas would be achieved through the use of a "Common Mining Code". Administrative uniformity in the coastal waters and the adjacent area off an individual state would be facilitated through the state minister's appointment as the Designated Authority responsible for handling

---

101. *State Powers Act, id.*
102. *Petroleum (Submerged Lands) Amendment Act 1980* (Cth) (No.80). It is of some interest to note that the 1980 amendments were not proclaimed until February 1983. *See Commonwealth of Australia Gazette*, No. S29, Wednesday, February 9, 1983.
103. *Id.*, ss.6 and 7.

day-to-day matters in the adjacent area. The Commonwealth's dominant role in each Joint Authority would ensure the pursuit of common policies in all adjacent areas. The extent to which the system has remained uniform will be discussed later.

### Decision-Making Mechanisms

As intimated above, all the legislation that governs offshore petroleum activities is similar: the Commonwealth adjacent areas are governed by the Commonwealth PSLA, while each state's coastal waters are governed by state legislation that is modelled on the Commonwealth PSLA. Little more will be said at this point about the state offshore petroleum legislation, except to note that the sort of powers that are vested in the Designated Authority and the Joint Authority under the Commonwealth PSLA typically lie with the state minister under the state legislation.

Under the Commonwealth PSLA, decision-making powers are shared between the Designated Authority and the Joint Authority. Table E contains a list of powers vested in the Joint Authority, while Table F contains a list of the powers of the Designated Authority.

The term "Designated Authority" [hereinafter DA] is defined in s.5(1) of the PSLA; the result of this definition, the definition of "State Minister" in the same section, and s.14 (entitled "Designated Authority") is that the DA is the relevant state minister.

Part 1A of the PSLA, added in the amendments which gave effect to the *1979 OCS*,[104] deals with the Joint Authority [hereinafter JA]. A JA, consisting of the Commonwealth minister and the relevant state minister, is established in respect of each state adjacent area. The JA's business may be conducted in writing or at meetings. Where there is disagreement as to a decision, or where the state minister has not replied within thirty days to a written notice containing the views of the Commonwealth minister, s.8D(2) entitles the Commonwealth minister to take the necessary

---

104. *Supra*, note 102, s.10.

decision:

> If the members of a Joint Authority disagree with respect to the decision to be made on a matter within the functions of the Joint Authority or the State Minister ... has not stated to the Commonwealth Minister his opinion as to the decision to be made on such a matter after having been given by the Commonwealth Minister not less than 30 days notice in writing of the opinion of the Commonwealth Minister as to the decision that should be made on the matter, the Commonwealth Minister may decide the matter ....

The interpretation of s.8D(2) has received judicial comment in a case that is dealt with in detail below;[105] for present purposes it may be noted that, in the case, the Victorian minister sought judicial review of a direction given to him by the Commonwealth minister. The DA for Victoria had given licensees notice of his intention to renegotiate the manner of royalty calculation. He copied this letter to the Commonwealth minister who, more than thirty days before making the decision being reviewed, gave the Victorian minister his written opinion on the matter without indicating when he would take a decision in the event of disagreement or the Victorian minister's failure to reply.

The Victorian minister asserted that s.8D(2) requires the Commonwealth minister to notify the state minister as to the time period within which a decision will be taken in the event of disagreement. Although admitting that this argument had some weight, the Federal Court rejected it:

> But this requirement of notice is for the regulation of the proceedings of agents of eminent status of executive governments who are assisted in the performance of their functions by a numerous and skilled staff. Those considerations are here lacking which might move a court to construe a statutory provision in similar idiom for the regulation of private contractual or proprietary rights as requiring specification of the period to be allowed. The words of the requirement as to notice are satisfied, literally, by the lapse of the specific period of 30 days between giving of the notice and the taking of the decision: after the lapse of that period the State Minister has been given not less than 30 days notice in writing of the Commonwealth Minister's opinion.[106]

---

105. See text accompanying note 121, *infra*. *Fordham and the State of Victoria* v. *Evans* (13 November 1987), No. VG428 of 1986 (Aust. F.C.).
106. *Id.*, at 8.

In 1983, the Commonwealth issued Guidelines pertaining to the handling of JA matters[107] intended to "ensure a uniform and consistent approach" to the administration of the PSLA. Applications concerning matters to be decided by the JA are to be forwarded by the state Department of Mines to the Petroleum Division in the Commonwealth Department.[108] The Commonwealth is to be advised of state views and vice versa. In the event that views differ, officials will attempt to resolve the matter by consensus. The state minister will then send his written opinion to the Commonwealth minister. Once a decision is reached, the applicants will be so advised by the state minister.

As one might expect, most JA matters are handled through an exchange of paper between the Commonwealth and the state departments. Although there have been some disputes, most matters have been resolved by discussion. Procedures have been streamlined and urgent matters can be expedited, but the process can still take six weeks or longer. Concern about delays in the JA decision-making system is expressed by both state officials and industry representatives. State complaints to this effect may reflect dissatisfaction about the growing Commonwealth role (discussed below) and the resulting loss of power at the state level. Some state officials explain certain delays on the basis that the Commonwealth time-frame is influenced by national considerations (for example, the desirability of announcing the award of several offshore permits at the same time); such considerations did not come into play under the *1967 Agreement* when the states had most of the administrative control and do not tend to arise under the *1979 OCS* in relation to matters within DA jurisdiction.

---

107. Australia, Dept. of Resources and Energy, *Guidelines for Handling Matters Requiring Commonwealth and State/Northern Territory Decisions, 83/1.*
108. In 1989 the relevant Commonwealth Department was called the Department of Primary Industries and Energy; because it has undergone a number of name changes over the past decade, it is hereinafter referred to simply as the Commonwealth Department.

From an industry perspective, DA decisions can be procured with relative ease, as the state department is on site and communications are frequent and mostly smooth. The JA procedure is perceived to be less clear-cut, however, requiring constant sensitivity to intergovernmental politics. The involvement of both levels of government can sometimes be advantageous to industry in that one level of government may take up industry's point of view in debates with the other. On the other hand, each level of government can blame delays on the other, leaving industry caught in the middle. Commonwealth officials reply to charges of delay by pointing out that the JA's decisions are often complex, requiring lengthy negotiations and long lead times.

Because there are so many steps in the JA process, opportunities for miscommunication abound. Part of this difficulty may lie in the structure of the Commonwealth Department. The Department refers certain matters to one of its divisions, the Bureau of Mineral Resources [hereinafter BMR]. Because there is no direct interaction between the BMR and state officials, the BMR may receive incomplete information from the Commonwealth Department and recommend inappropriate action.

As mentioned above, the PSLA has been amended on several occasions since 1983 in ways that decrease the authority of the DA and increase that of the JA. This may be viewed as a fine-tuning of the system or, on the other hand, as growing Commonwealth control over the adjacent areas. A few examples of such changes demonstrate the direction that is being followed.

In 1984 the power to vary the rate of petroleum production was transferred from the DA to the JA. The resulting s.58(5) specifies that, in giving the licensee directions as to petroleum production, "the Joint Authority may take into account matters relating to the effects on Commonwealth revenue of the proposed direction". A ministerial speech at the time of the amendment explained:[109]

---

109. *Petroleum (Submerged Lands) Amendment Bill 1984* (Cth) (No.166): Second Reading, Speech By The Minister Representing The Minister For

> This will remove an inconsistency in the present legislation whereby the Joint Authority is responsible for directing commencement of production but the Designated Authority is empowered to direct licensee to increase or reduce that production.

In fact, it appears that this amendment resulted from a dispute between the Commonwealth and Victoria over the appropriate rate of production from the Fortescue field.[110] The DA had given permission for oil from the field to be produced at high levels, a situation that maximized the state royalty at the expense of the Commonwealth excise tax. The Commonwealth response was to transfer power over production control from the DA to the JA.

Amendments to the PSLA passed late in 1987 take further this trend toward centralized power. They accomplish the following:

- Remove Western Australia's special decision-making rights as formerly contained in s.8D(9).
- Repeal s.8E, pursuant to which a list of matters contained in a schedule to the PSLA were for DA decision-making *unless* the Commonwealth minister notified the DA that these matters were for the JA. In the future, such matters will automatically be subject to JA determination.
- Transfer from the DA to the JA powers under ss.36 and 37 pertaining to the declaration of a location.[111]
- Transfer from the DA to the JA most of the powers under s.103 to vary, suspend, or exempt from compliance with the conditions of various title documents.
- Permit the Commonwealth minister to give directions to the DA pursuant to s.122 as to what accounts, records, and other documents should be kept and furnished by persons carrying on operations in the adjacent area.

---

Resources and Energy, at 3.

110. Australia, Dept. of the Auditor-General, *Efficiency Audit Report: Department of Resources and Energy: Offshore Petroleum Royalties and Excise* (Canberra: Australian Government Publishing Service, 1987) [hereinafter *Auditor-General's Report*], at 42-43.

111. The location status pertains to blocks where petroleum has been discovered; the declaration of a location provides an intermediate step between exploration and production.

The reasons for these changes were explained by the Commonwealth minister at the time of the Bill's second reading.[112] In late 1985, the Labour Government's review of the *1979 OCS* had led to the conclusion that the regime was operating satisfactorily but that certain changes were needed. The Commonwealth government felt that all states should be treated the same and that Western Australia's special rights should be removed. The repeal of s.8E and changes to s.103 reflected the fact that, in practice, these matters were determined by the JA rather than the DA:

> This practice has evolved in order to avoid the significant administrative delays which would otherwise occur if the Commonwealth Minister was to decide that a matter should be referred to the Joint Authority ... A further consideration in the Government's decision to refer all these matters to the Joint Authority has been the apparent confusion, particularly for new companies wishing to undertake offshore petroleum operations, about which authority is responsible for decisions on such matters.[113]

The additions to s.122 flowed from a review of procedures concerning verification of royalty collections by the Australian Auditor-General's office and were intended to clarify the Commonwealth's right to obtain documents related to offshore production.[114]

Further diminution of the DA's powers resulted from the *Petroleum Resource Rent Tax (Miscellaneous Provisions) Act 1987*,[115] passed on December 15, 1987. These amendments transferred from the DA to the JA the power (contained in s.6A(4)) to determine the amount of petroleum to be assigned from a pool which is contained in more than one licence area and powers (contained in s.59) to make directions concerning unitization.

After relatively short experience with the *1979 OCS*, therefore,

---

112. *Petroleum (Submerged Lands) Amendment Bill 1984* (Cth) (No.166): Second Reading, Speech by The Hon. John Kerin, M.P., Minister for Primary Industries and Energy.
113. *Id.*, at 3.
114. *Id.*, at 8.
115. *Petroleum Resource Rent Tax (Miscellaneous Provisions) Act 1987* (Cth) (No.145).

the Commonwealth has found it desirable to consolidate certain powers at the national level. Because the PSLA is a Commonwealth statute and the 1975 *Seas and Submerged Lands Case* decision clearly gives the Commonwealth legislative authority over the continental shelf, the amendment process has been straightforward. The states have been consulted about the changes, and while not necessarily in favour of them, have not been prepared to treat them as a major political issue.

Some state officials resent the growing Commonwealth role, stating that the JA's increasing power makes it more and more difficult to regulate offshore petroleum arrangements efficiently and expeditiously. There is some state sentiment to the effect that the Commonwealth is crossing the line between policy and operations, ignoring the fact that the *1979 OSC* was designed to leave operational matters in state hands. At the same time, there is grudging recognition that the PSLA is Commonwealth legislation which the Commonwealth is legally entitled to alter as it sees fit. State officials also admit that many of the changes relate, directly or indirectly, to the raising of Commonwealth revenue, a matter of legitimate Commonwealth concern.

## Legislative Authority

The division of legislative authority pertaining to the management of Australia's offshore petroleum resources is different than that found in Canada. The complexity of this topic, however, makes it worthwhile to summarize briefly the Australian legislative regime.

Section 5(a) of the *State Powers Act* provides that the legislative powers exercisable under the Constitution of each state extend to the making of all such state laws as could be made if the coastal waters of the state were within state limits. These legislative powers are limited by s.7(c) which provides:

> Nothing in this Act shall be taken to ... give any force or effect to a provision of a law of a State to the extent of any inconsistency with a law of the Commonwealth or with the Constitution of the

> Commonwealth of Australia or the Commonwealth of Australia *Constitution Act.*

These provisions enable the states to apply their petroleum and other laws to their coastal waters but to retain the general design of the Constitution as to supremacy of Commonwealth laws relating to matters within Commonwealth jurisdiction (such as taxation).

Provisions concerning legislative authority over the adjacent areas (i.e., beyond the three-mile coastal waters) are contained in Part II of the PSLA. Section 9(1) of the PSLA provides:

> Subject to this Act and the regulations, the provisions of the laws, whether written or unwritten, in force in a State for the time being (other than laws of the Commonwealth) and the provisions of any instrument having effect under any of those laws, apply, as provided by this section, in the adjacent area and so apply as if that area were part of that State and of the Commonwealth.

As a result of this section and s.9(4) and (5), the PSLA itself governs the management of petroleum resources in the adjacent areas, but much of the resulting activity is regulated by state law (such as that pertaining to occupational health and safety).

The general recognition of state legislative authority over the adjacent area contained in s.9(1) is specifically limited by other parts of s.9. For example, s.9(6)(c) and (d) state that s.9 does not extend any law in such a way that it would be inconsistent with a Commonwealth law or apply so as to impose any tax. Perhaps most importantly, s.9(8) enables state law to be made inapplicable to an adjacent area by virtue of regulations made under the PSLA. Although this power does not appear to have been exercised, it gives the Commonwealth an effective tool with which to limit the applicability of state law within an adjacent area, should that prove desirable. The states have responded in varied ways to their offshore legislative authority, an explanation of which can be found elsewhere.[116]

## Fiscal Arrangements

Four clauses in the *1967 Agreement* dealt with fiscal

116. *See* Cullen, *supra*, note 89, at 123-136.

arrangements. Section 19 provided that a royalty of up to ten per cent of the "value at the well-head of the petroleum" would be divided on a 60/40 per cent basis between the relevant state and the Commonwealth, while any additional "over-ride" royalty would be allocated to the state. Section 20 stated that royalty reductions could be made by the DA only on agreement between the Commonwealth and state ministers and s.21 required consultation between the DA and the Commonwealth minister before the DA exercised the power to determine the value of any petroleum, otherwise than by agreement with the permittee or licensee. Section 22 gave the states the benefit of "all moneys, other than royalties, payable under the Common Mining Code".

Although these principles were scrutinized by the Senate Select Committee, it had only been asked to consider whether proper provision had been made in the legislation for adequate royalties used in the national interest. The Committee noted the existence of problems pertaining to the calculation of "wellhead value" and pointed to the need for common rules in the assessment of transport and treatment costs, but offered no critical comments on the overall *1967 Agreement* fiscal regime.

The material released pertaining to the *1979 OCS* contains only one statement concerning fiscal matters: "The present arrangements for the sharing of royalties between the States and the Commonwealth will be retained."[117] Alterations were made to the *Petroleum (Submerged Lands) Royalty Act 1967* [hereinafter *Royalty Act*] in 1980 as part of the process of implementing the *1979 OCS*. For example, it was made clear that money received by a DA in relation to petroleum activities in an adjacent area would be received on behalf of the Commonwealth and the JA was given supervisory powers over the DA in relation to royalty-related matters.[118]

---

117. *Offshore Constitutional Settlement*, *supra*, note 94, at 8.
118. Sections 5(1B) and 12(2C) of the *Royalty Act*, *supra*, note 82, provides that moneys are received by the DA on behalf of the Commonwealth, while s.13 makes certain sums payable by permittees and licensees "debts due ... to the Commonwealth". Sections 10A and 12(2A) authorize the JA

The revenue-sharing arrangements of the *1979 OCS* are deceptively simple as there is probably no other aspect of the offshore arrangements that has given rise to more problems. Historical difficulties in determining the "wellhead value" have led to High Court litigation between the Victorian DA and certain licensees.[119] Of greater interest here are the three problems that have emerged at the Commonwealth-state level: the treatment of the crude oil levy for royalty calculation purposes, Commonwealth involvement in royalty calculations, and the Resource Rent Tax.

### *Treatment of the Crude Oil Levy for Royalty Calculation Purposes*

Following the rapid escalation of oil prices in 1973-74, most nations took action to capture the "windfall profits" that accrued to oil companies. In Australia, the major tool was the excise levy on crude oil imposed by the Commonwealth in 1975.[120] Inevitably, it had to be asked how this crude oil levy would be treated in the calculation of royalty: if treated as a deduction prior to the calculation of royalty, it would dramatically reduce the state's royalty revenue while increasing the Commonwealth take.

Section 9 of the *Royalty Act* provides for agreement to be reached between a DA and a licensee as to "the value at the well-head of any petroleum". In an exchange of letters in June 1980 between the Victorian DA and BHP Petroleum and Esso concerning royalty calculation in the Gippsland Basin, the Victorian DA agreed that "duties of customs and excise will not be

---

to give the DA directions in relation to the exercise of certain DA powers.

119. In *BHP Petroleum Pty. Ltd.* v. *Balfour*, [1987] 61 A.L.J.R. 345 (Aust. H.C.), the High Court ruled that the DA's power to determine the location of the "well-head" cannot be exercised in an arbitrary fashion. For a comment on the case, *see* Garrie Moloney, "Case Notes: *BHP Petroleum Pty. Ltd.* v. *Balfour*" (1987) 16 M.U.L.R. 436.

120. A brief description of the background to the crude oil levy is found in Australia, Commonwealth Parliament, Joint Committee of Public Accounts, *190th Report: Inquiry into Petroleum Royalties and Excise*, Parliamentary Paper No. 236/1981 (Canberra: Australian Government Publishing Service, 1981) at 9.

included in the assessment of wellhead value".[121] It is not known why the Victorian government took this position and accepted the resulting loss of royalty revenue. Some suggest that while the Victorian government always resented the crude oil levy's impact on its royalty revenues, the Commonwealth convinced it that national revenue-raising had to take precedence over state revenue-raising through offshore oil royalties. In the intervening years, the state government unsuccessfully sought other means of generating revenue from offshore activities.[122]

In a letter to the licensees dated September 29, 1986, the Victorian DA gave notice of revocation of the earlier agreement as to wellhead calculation. The Victorian DA supplied a copy of this letter to the Commonwealth minister, who, on November 11, 1986, directed the DA to seek to reach a new agreement with BHP and Esso within three days, by using the same method of calculation as that contained in the June 1980 exchange of letters. In default of reaching such an agreement, the Commonwealth minister directed the DA to employ the same calculation method as had been agreed in the June 1980 exchange of letters.

The State of Victoria sought review of the direction in the Federal Court pursuant to the *Administrative Decisions (Judicial Review) Act 1977*.[123] Earlier reference has been made to one of the grounds for review, the interpretation of s.8D(2) of the PSLA.[124] A second ground concerned the interpretation of s.10A(1) of the *Royalty Act*, which provides:

> The Joint Authority shall give directions to the Designated Authority *with respect to the manner in which the Designated Authority is to exercise his powers* under sections 8, 9 and 10 and the Designated Authority shall not exercise any of those powers except in accordance

---

121. The letter is quoted in *Fordham* v. *Evans*, *supra*, note 105, at 13.
122. For example, by imposing a licence fee on pipelines transporting petroleum from the Gippsland Basin. *Hematite Petroleum Pty. Ltd.* v. *State of Victoria* (1983), 151 C.L.R. 599 (Aust. H.C.), discussed in Chris Caleo, "Section 90 and Excise Duties: A Crisis of Interpretation" (1987) 16 M.U.L.R. 296.
123. *Administrative Decisions (Judicial Review) Act 1977* (No.59).
124. See text accompanying note 105, *supra*.

> with directions so given. [emphasis added]

Sections 8, 9, and 10 pertain to the DA's power to ascertain the wellhead location, the value of petroleum at wellhead, and the quantity of petroleum recovered. Victoria asserted that the JA's power in s.10A related only to the procedure and not to the substance of the calculation or the exercise by the DA of his authority. Relying in part upon the view of the DA's s.8 authority taken by the High Court in *BHP Petroleum Pty. Ltd.* v. *Balfour*,[125] Jenkinson J. concluded:

> I see no reason to confine the meaning of the word "manner" to the sense for which the applicants contend. Normal usage of the word in the context which that sub-section and those three sections supply would comprehend matter of substance as well as matter of procedure.[126]

The state had argued that, as this interpretation would leave no substantial function to be exercised by the DA, it would abrogate the concept of "sharing" between the states and Commonwealth referred to in one of the recitals to the PSLA. This argument was given short shrift by Jenkinson J. who stated: "But it was not shown that any such agreement assured to the State or Territory any particular administrative function".[127]

Since this decision, Victoria has launched a new action attacking the validity of the crude oil levy itself,[128] apparently as part of its overall negotiating strategy in relation to Commonwealth -state revenue issues. Whatever the outcome, this thorny question reflects longstanding state dissatisfaction with the results of offshore petroleum revenue-sharing.

### *Commonwealth Involvement in Royalty Calculations*

The role of the Commonwealth government in assessment and verification of offshore royalties has proven to be another problem.

---

125. *BHP Petroleum Pty. Ltd.*, *supra*, note 119, at 17.
126. *Fordham*, *supra*, note 105, at 17.
127. *Id.*
128. The case and offshore revenue-sharing in general are discussed in Richard Cullen, "Bass Strait Revenue Raising: A Case of One Government Too Many?" (1988) 6 J.E.N.R.P. 213.

Because of references to the issue in the Commonwealth Auditor -General's reports from 1977 to 1981, the Parliament's Joint Committee of Public Accounts decided to examine the subject in its 1981 report.[129] Among the matters raised by the Auditor -General were whether the procedures followed in the accounting for and sharing of offshore royalties complied with the PSLA and whether the Commonwealth Department had adequate involvement with Victorian officials in verifying royalties.[130]

The Joint Committee took a somewhat dim view of the *1979 OCS*, in particular expressing doubt as to whether the OCS adequately protected the Commonwealth's position with respect to the collection of royalties:

> Because of the manner in which certain administrative functions in relation to the territorial sea and the adjacent areas are passed over to the designated authority the Committee considers that in a practical sense the position of the Commonwealth in relation to the collection of money is not fully safeguarded. Should the Commonwealth wish to inquire into whether royalties are being properly collected there could be difficulties as it is likely that those with the day to day responsibility for the collection of the Commonwealth's share of royalties will be State Government officials and therefore under no obligation to furnish full explanations to the Commonwealth Parliament.[131]

Another problem the Joint Committee identified was that the state's major role in calculating royalties could conflict with Commonwealth interests in regard to the collection of the crude oil levy. Specifically, crude oil might be allocated to a field which attracted a high over-riding royalty but a low rate of excise levy.[132] The Joint Committee concluded that the Commonwealth Department had been "too willing to vacate the field in favour of the State authorities and has not made sufficient effort to establish check procedures by which to ensure that royalty payments are

129. *Inquiry into Petroleum Royalties and Excise*, *supra*, note 120, at v.
130. *Id.*, at c.4.
131. *Id.*, at 15-16.
132. *Id.*, at 29.

correctly determined".[133]

Some six years later, the Commonwealth Auditor-General undertook an efficiency audit of the Commonwealth Department (then the Department of Resources and Energy [hereinafter DRE]) to "assess whether DRE obtained adequate assurance that the correct amount of royalty had been collected".[134] Among other things, the Auditor-General concluded that DRE had not fulfilled its responsibilities in the administration of the offshore oilfields, having been slow to accept its role and being unable to provide satisfactory evidence that all due revenue had been collected.[135] DRE's response was that the *1979 OCS* was complex administratively and that:

> Audit has not given due recognition to the roles of the Commonwealth and the States (in this case Victoria) in this area and consequently has provided an overly simplistic representation of what is an exceedingly complex area.[136]

More specifically, the Auditor-General expressed doubt that the legislation, as then structured, would entitle DRE to have sufficient access to Victoria's and the producers' accounting systems to conduct *ad hoc* reviews for the purposes of verifying the royalties.[137] His recommendation was that there be an assessment of DRE's legal position in this regard, which has resulted in amendments to the PSLA.[138]

Another matter noted by the Auditor-General concerned a direction issued by the Victorian DA in relation to approval for Esso's method of production measurement.[139] The Auditor-General suggested that this approval should have been given pursuant to s.10 of the *Royalty Act* rather than pursuant to a

---

133. *Id.*, at 31.
134. *Auditor-General's Report*, *supra*, note 110, at 11. As mentioned at 3 of the *Report*, efficiency audits "involve an evaluation of the effectiveness of administrative action and decisions taken by officials in achieving program objectives".
135. *Id.*, at 1.
136. *Id.*
137. *Id.*, at 17.
138. See text accompanying note 114, *supra*.
139. *Auditor-General's Report*, *supra*, note 110, at 39.

provision in the "Directions as to Petroleum Production". In the former case, the DA's direction would have been subjected to JA supervision, whereas in the latter case it would not. The Auditor-General recommended that Commonwealth-state discussions be initiated to determine when the "Directions as to Petroleum Production" should be used in such a situation rather than directions by the DA that would be subject to JA approval.

Although the Auditor-General makes no suggestion that the DA had chosen this method deliberately to avoid JA supervision, the situation highlights a general issue concerning the use of directions. As this matter is explored more fully below,[140] it is sufficient to note here that s.101(1) of the PSLA gives the DA broad powers to issue directions to licensees. In theory, the DA's power is circumscribed by s.101(3), which states that the DA shall not give a direction "of a standing or permanent nature except with the approval of the Joint Authority". In light of some of the information contained in the *Auditor-General's Report* and the reluctant state acceptance of the Commonwealth's growing presence in the adjacent area, it is not unimaginable that the restriction in s.101(3) is less honoured in practice than on paper.

Some producers have expressed concern about certain aspects of the *Auditor-General's Report*, in particular about costs and inefficiencies that could result from "the proliferation of uncoordinated independent audits" of operations.[141] The Commonwealth's continuing concern about royalty calculation problems is reflected in the fact that it has tendered for an outside study of the issue.

### *The Resource Rent Tax*

Discussion about the benefits of a profits-based royalty or tax on petroleum has taken place in Australia since 1976.[142] The

---

140. See note 173 and accompany text, *infra*.
141. *Auditor-General's Report*, *supra*, note 110, at 3.
142. Background to this concept is contained in Varis Vitols, "Petroleum Taxation in Australia: The Resource Rent Tax Bills 1986" (1987)

notion behind this concept is that, while the public is entitled to receive benefits from community-owned resources (such as petroleum), producers should be guaranteed a return on their investment. In 1983, the Labour Government circulated a discussion paper concerning the concept of a "Resource Rent Tax" [hereinafter RRT] to replace both the crude oil levy and royalty charges on petroleum production.

A RRT has now been introduced for Commonwealth adjacent area petroleum projects for which a production licence had not been obtained as of July 1, 1984, and which are not located in permit areas from which, as of the same date, licences were drawn.[143] In these so-called "greenfields" areas, no crude oil levy or royalty will be applied. The first field to produce under the new scheme is Jabiru in the Timor Sea, an area that is part of a Commonwealth territory.

The Commonwealth government had hoped to apply the same concept to projects under state authority, but problems in achieving agreement with the states have necessitated a different treatment of such projects. In relation to those projects, the Commonwealth's *Petroleum Revenue Act 1985*[144] contemplates an agreement between a state and the producers of a project subject to a Resource Rent Royalty [hereinafter RRR], as well as a revenue-sharing agreement between the state and the Commonwealth. This mechanism has been used to renegotiate the *Barrow Island Agreement* pertaining to production on Barrow Island, an area under Western Australia's jurisdiction. Under the previous Commonwealth crude oil levy and state royalty regime, Barrow Island production would have been marginal. Under the *Petroleum Revenue Act 1985*, production will only be subject to the RRR, similar to the RRT, which will be shared 75/25 between the Commonwealth and Western Australia. It is hoped that this regime

---

11 Adel. L.R. 140. The article analyzes the RRT in considerable detail.

143. The information which follows is taken from *Energy 2000*, *supra*, note 98, at 10 *ff*. The measure will be handled through a series of statutes outlined in Vitols, *id*.

144. *Petroleum Revenue Act 1985* (Cth) (No.188).

will encourage exploration and development on Barrow Island.

The Commonwealth sees the *Barrow Island Agreement* as a model for projects under state control. A report prepared for the Western Australian government, however, suggests that the *Barrow Island Agreement* is not necessarily a beneficial precedent for that state. Pointing out that division of royalties between the Commonwealth and state has "been a matter of the relative power of the two levels of government, with the Commonwealth being both the more powerful and the more aggressive",[145] the report's author objects to the new scheme because a rising Commonwealth share will result in a diminishing state share and because the new system interferes with state resource management by restricting the state's freedom to choose a royalty rate and definition of royalty base. While recognizing that one objective of the new Commonwealth system is the harmonization of state policies, he is not convinced of the desirability of this goal: "Thus Western Australia differs from, for example, Victoria in its stage of development and its resource endowment. It may wish to manage its resources to secure a different mix of objectives."[146]

The report recommends:

> that the State in its dealings with the Commonwealth place a high priority on maintaining the authority it needs to properly manage the mineral resources to which it has been granted title. Moreover, the State should not compromise its claim to royalties, inasmuch as the prospect of royalty income is an incentive to efficient resource management.[147]

The Commonwealth's experience in attempting to simplify fiscal aspects of petroleum exploitation demonstrates how new policy initiatives must always be examined against the backdrop of federal-state relations. The *Seas and Submerged Lands Case* undoubtedly gives the Commonwealth the constitutional authority to revise the fiscal regime in the adjacent area, but it is unclear how

---

145. Western Australia, Mineral Revenues Inquiry, *Final Report: The Study into Mineral (Including Petroleum) Revenues in Western Australia* (Paul G. Bradley, Director) vol.1 (1986), at 214.
146. *Id.*, at 215.
147. *Id.*, at 240.

the RRT will affect the royalty-sharing arrangements agreed in the *1967 Agreement* and affirmed under the *1979 OCS*. The division of the RRT between the two levels of government is a matter under negotiation and is seen by some as an important test of the OCS. Whether the RRT will be applied to state coastal waters remains an open issue. The Commonwealth government would have liked to apply the RRT to the entire offshore, but was reluctant to do so in the face of state objections for fear such action would be seen as breaching the OCS's principle of state control over coastal waters. Ironically, it does not seem to have occurred to Commonwealth officials that the imposition of the RRT, in the absence of agreement as to how it would be shared, might also be seen as a breach of the OCS.

### *Summary*

The above analysis reveals that fiscal matters have continued to be a Commonwealth-state irritant under Australia's offshore petroleum regime. The treatment of the Commonwealth crude oil levy for royalty calculation purposes, the role of the Commonwealth in royalty calculation and verification, and the sharing of the new RRT remain unresolved problems. As will be discussed in the section entitled "Uniformity Between the Commonwealth PSLA and State Offshore Petroleum Laws", entitlement to proceeds from the new cash bidding system looms as another issue. While Commonwealth Guidelines concerning the collection and verification of royalties, fees and other moneys have been in existence since 1983,[148] underlying questions concerning the division of revenues between the two levels of government will not go away. These difficulties are undoubtedly part of the larger problems of Commonwealth-state revenue-sharing that

---

148. Australia, Dept. of Resources and Energy, *Offshore Petroleum Production Beyond the Three Mile Territorial Sea — Procedures for the Collection and Verification of Royalty, Fees and Other Moneys, 83/4.*

characterize relations under the Australian Constitution.[149] Nevertheless, the persistence of these issues in the offshore context inevitably influences the smooth operation of the system and has an impact upon the industry.[150]

## Boundary Setting and Resolution

Although the *1967 Agreement* was silent as to the procedure that would be used to divide the offshore between states and federal territories, the PSLA contained schedules that delimited the various areas. The setting of these boundaries "required delicate negotiation and was one of the last matters to be resolved".[151] Problems arose in the determination of the boundary between Victoria and South Australia, but eventually agreement was reached.[152]

Following the *1979 OCS*, another set of boundaries had to be drawn to distinguish between the three-mile territorial sea (the "state coastal waters") and the Commonwealth adjacent area beyond that would be governed by the PSLA. The OCS had specified that the state areas would be limited to three miles regardless of whether Australia later moved to a twelve-mile territorial sea, a principle reflected in s.5A of the PSLA. The OCS material also stated that:

> [T]he baselines from which the territorial sea will be measured will be drawn in a way that takes advantage of the international principles authorizing the drawing of "straight baselines" where the coast is deeply indented or fringed by islands, and of closing lands where bays are not more than 24 miles wide ... The "internal waters" on the

---

149. Australia's overall system of revenue-sharing was criticized by Mr. John Cain, Premier of Victoria, in "Reform of Federal Finance Seminar" (Address at the University of Melbourne, 24 February 1988) [unpublished].
150. While it is impossible to quantify what impact these issues have had on industry, it is clear that, at the least, they involve a loss of executive time due to participation in activities such as the Auditor-General's inquiries and litigation such as Victoria's actions against the Commonwealth.
151. C.A. Wells, *Co-operative Federalism in Australia* (Ph.D. thesis, University of Melbourne, 30 June 1976) [unpublished] at 353-54.
152. Described *id.*

> landward side of these lines will be included in the grants made by the legislation. The result will be to enlarge the area in which the States will enjoy the benefits of the legislation.
>
> The baselines to be adopted are being prepared in close consultation with the States and will be promulgated in due course under the *Seas and Submerged Lands Act 1973*.[153]

Since the states saw the drawing of the baselines as their final chance to maximize the areas under their jurisdiction, the negotiations were time-consuming and some states took a hard line concerning what principles should be used and how they should be applied. Rather than utilizing the mean low water-mark, the Lowest Astronomical Tide (the lowest level that can be predicted to occur under average meteorological conditions) was employed to draw the baselines.[154] Once the baselines were drawn, it was a relatively straightforward task to measure the three-mile territorial sea.[155]

With two sets of offshore boundaries (between the states, and between the state's coastal waters and the Commonwealth adjacent areas), the PSLA has to make provision for a variety of issues that might arise, including the following situations:

- Petroleum pools spanning more than one adjacent area (s.6A(3))
- Petroleum pools spanning state coastal waters and a Commonwealth adjacent area (s.6A(5))
- Unit developments across boundaries (s.59)
- Access for exploration purposes to a state area from a Commonwealth adjacent area (s.112(A))

In regard to petroleum pools that cross these boundaries, s.6A(3) and (5) of the PSLA provide that there shall be deemed to

---

153. *Offshore Constitutional Settlement*, *supra*, note 94, at 7.
154. Details of the procedures followed are contained *in* Australia, Dept. of the Attorney General, *Australia's Territorial Sea Baseline* (Canberra: Australian Government Publishing Service, 1988). The baselines were declared through Proclamations by the Governor-General in 1983 and 1987.
155. State jurisdiction extends to offshore islands. There appears to be one area where jurisdiction is still unresolved, namely, the Scott Reef, a sandy area off Western Australia whose shape changes with the tides.

have been recovered in respect to each area "such proportion of all petroleum so recovered as may reasonably be treated as being derived from that area, having regard to the nature and probable extent of the pool". The relative proportions are to be determined in accordance with s.6A(4) and (6). Until the 1987 PSLA amendments, proportions relating to pools crossing adjacent area boundaries could be determined by agreement between the licensee and the two DAs, or by the Supreme Court of a state upon application by the licensee or one of the DAs. The DA's former power in these regards now rests with the JA. Where the pool crosses the boundary between a state's coastal waters and a Commonwealth adjacent area, the procedure is similar except that the relevant authorities are the state minister and the JA.

Section 6A(7) provides for the situation where a Unit Development Agreement is in place, in which case the proportions will be determined in accordance with that agreement. The DA's former powers under s.59 in relation to Unit Development Agreements were also transferred to the JA in 1987. Although s.59 gives a JA broad authority to give a licensee directions concerning unit development and to approve Unit Development Agreements, where boundaries between state coastal waters are involved, the DA is obliged by s.59(11) to consult with any other DA concerned or with the appropriate authority of another state. Section 59(12) provides that, in such a situation, the JA cannot give a direction or approve a Unit Development Agreement except with the approval of any other JA concerned and any other state authority concerned. Presumably, if such approval cannot be obtained, resort can be taken to the judicial resolution provided for by s.6A(4) and (6).

Despite these provisions in the PSLA, other transboundary problems are likely to arise. At least one prospective area has already required careful intergovernmental handling, namely, an area spanning Victoria's coastal waters and the Commonwealth adjacent area. Issues that had to be addressed included which licensing regime would apply (specifically, whether the Commonwealth's cash bidding procedure, discussed in the section entitled "Uniformity Between the Commonwealth PSLA and State Offshore Petroleum Laws", would be utilized even though it has no counterpart in Victoria's legislation) and what royalty regime would be utilized (the new Commonwealth RRT or Victoria's royalty system). Negotiations between the two governments eventually led to agreement that the RRT would apply along with a modified form of cash bidding.[156]

Transboundary issues have caused many practical and legal problems offshore Western Australia, where it is possible to have as many as three petroleum management regimes applying to one prospective area.[157] Legal advisers point to the difficulties of searching titles under three registry systems and of coping with a multiplicity of reference names and numbers. If a transaction is governed by more than one regime, fees have to be paid pursuant to each even though, as between the parties, there is only one consideration. Another frustration is the fact that a transaction must be registered under each applicable regime. With the prospect of new petroleum areas that may span the various boundaries, concern is expressed about possible economic unpredictability if it is not known whether the state or Commonwealth fiscal regime, or a combination thereof, will apply.

---

156. The transboundary area contains the Golden Beach gas discovery. *See* Australia, Minister for Resources and Energy, Press Statement 99/86, 16 September 1986. The cash bidding issue is discussed in the section entitled "Uniformity Between the Commonwealth PSLA and State Offshore Petroleum Laws".
157. Activities in Western Australia's internal waters are governed by the state's onshore petroleum legislation (Petroleum Act), in the coastal waters by the state PSLA-counterpart, and in the adjacent area by the Commonwealth PSLA. This complexity is exacerbated by the arrangements that governed the transition from the *1967 Agreement* to the *1979 OCS*. For a detailed discussion, *see* Andrew G. Thompson, "Legal Aspects of Petroleum Tenement Management in Western Australia" *in* Australian Institute of Geoscientists, *Seminar No.4*, 22 September 1986, at 77.

## Selected Legal and Administrative Issues

### Amendment of the PSLA and Related Legislation

Two separate matters are discussed in this section: how amendments to offshore petroleum legislation (the PSLA and state-PSLAs) are procured, and the extent to which the Commonwealth may unilaterally undo the *1979 OCS*.

#### *Amendment of the PSLA and State Offshore Petroleum Laws*

The *1967 Agreement* was explicit concerning the method to be followed in amending the mirror state-Commonwealth legislation. Section 6(1) stated that no government would submit to its Parliament a Bill for an Act that would amend or repeal the mirror legislation or that would "in any material respect" affect the scheme of the legislation contemplated by the *1967 Agreement*, except in accordance with an agreement between the various governments. Section 7 extended this principle to the making, amending or repealing of regulations, while s.25(1) prohibited variation of the *1967 Agreement* itself without consensus among the governments.

While the PSLA was under discussion, this concept of multi-party consensus was criticized on the ground that the requirement of unanimity would result in the absence of amendments.[158] The Senate Select Committee, too, highlighted the problems that could arise, pointing out that the scheme had:

> inbuilt rigidity and depends for its adaptability to meet changing circumstances on a concurrence which can only be hoped to be forthcoming. Providing that the co-operation which marked the initial negotiations prevails, it is not unreasonable to assume that such amendments as experience suggests will be agreed upon. However, if relationships between the Commonwealth and the States are strained it is, equally, not unreasonable to assume that one State — or the Commonwealth — would withhold its concurrence as a bargaining consideration for the attainment of other objectives.[159]

The Committee itself identified one problem that had arisen

---

158. *Senate Select Committee Report*, *supra*, note 75, at 176.
159. *Id.*, at 177.

when, in March 1969, state-Commonwealth agreement was reached concerning the need to amend the mirror legislation.[160] The South Australia Parliament amended its legislation in late 1969, followed shortly by the Parliament of Western Australia. More than two and a half years after the two state governments had agreed to amend their statutes, however, neither the Commonwealth nor the four other states had taken the necessary action. Observing that, as a result, the legislative mirroring had been lost, the Senate Select Committee commented:

> Insofar as the constitutional conception underlying the legislation requires identical State and Commonwealth legislation, directions made, after the amendment, by the South Australia Designated Authority for example, to persons other than permittees, licensees, pipeline licensees or the holder of special prospecting authorities or access authorities, will lack the force which Commonwealth legislation, if it were to be similarly amended, may accord them. If the Commonwealth legislation should at any time be found to be the valid legislation, there must be real doubt as to the efficacy of acts done, rights accrued and the law applicable in South Australian offshore areas while the Commonwealth legislation remains unamended.[161]

Moreover, notwithstanding the requirements of s.6(1) of the *1967 Agreement*, it is not clear to what extent the affected states were consulted when, in 1973, the Commonwealth PSLA was amended to take account of three new boundaries between the Commonwealth and Indonesia and thereby altered the boundaries of Queensland, the Northern Territory, and Western Australia.[162]

The public documents concerning the *1979 OCS* make no reference to consultation or agreement over amendments, perhaps partly because of the difficulties that were experienced in procuring amendments under the *1967 Agreement.* The changed approach, of course, also reflects the fact that the *Seas and Submerged Lands Case* had validated Commonwealth jurisdiction over both the territorial sea and the continental shelf, making state involvement more a matter of policy than law.

---

160. *Id.*, at 180*ff.*
161. *Supra*, note 75, at 183.
162. This is described in Wells, *supra*, note 151, at 370*ff.*

Moreover, the scheme put into place under the *1979 OCS* is not of the "mirror" variety that characterized the *1967 Agreement.* Instead, the offshore has been divided into two main areas, one subject to the Commonwealth PSLA and the other to state offshore petroleum laws. This arrangement does not require identical legislative regimes for the legal reason that necessitated the mirror scheme in a period of constitutional uncertainty. Rather, as discussed in the section entitled "Consistency in Laws, Regulations, and Policies", a cooperative amending approach is desirable due to the goal of having uniform petroleum legislation throughout all offshore areas and has been achieved through the establishment of certain intergovernmental mechanisms.

***The Commonwealth's Power to Unilaterally Alter the 1979 OCS***

There has been considerable academic discussion in Australia concerning the Commonwealth's power to undo unilaterally the *1979 OCS* by amending or repealing the *State Powers* or *State Title* statutes. Because the complex legal arguments surrounding this issue turn largely upon unique features of Australian constitutional law and the unusual way in which the *1979 OCS* was legislated, the positions that have been put forward will be adverted to here only briefly.[163]

The first point to note is the way in which the *State Powers Act* was passed. Section 51(38) of the Constitution contemplates the exercise of power by the Commonwealth at the request or with the concurrence of the Parliaments of all the states directly concerned. Since this statute was passed following such state "request" legislation, it has been argued that state action is a prerequisite to its amendment or repeal by the Commonwealth. One counterargument commonly made is that since the

163. For a further discussion of the constitutional arguments, see Cullen, *supra*, note 89, at 96-100 and 104-106; Crommelin, *supra*, note 94; Kevin Booker, "Section 51 (xxxviii) of the Constitution" (1981) 4 U. of N.S.W. L.J. 91; R.D. Lumb, "Section 51, pl (xxxviii) of the Commonwealth Constitution" (1981) 55 Aust. L.J. 328.

Commonwealth had the legislative authority to pass this legislation, it implicitly has the legislative authority to amend or repeal it.

In relation to the *State Title Act*, a further complication results from s.51(31) of the Constitution, which gives the Commonwealth power to legislate with respect to "the acquisition of property on just terms from any State". It is widely agreed that this provision would require the Commonwealth to compensate the states if it were to diminish the title to the territorial sea which they have been granted under the *State Title Act*. If this view is correct, serious practical problems would face any Commonwealth government attempting to undo the legislation. A hint that this result was intended is found in a statement contained in the public material relating to the *1979 OCS*.[164]

It will be recalled that a Commonwealth Labour government passed the *Seas and Submerged Lands Act 1973* that led to the 1975 *Seas and Submerged Lands Case*. The Labour party had lost office before the *1979 OCS* was designed but regained power shortly after the legislation implementing the OCS was proclaimed in 1983. Since the Labour party had consistently opposed the notion of state power over the territorial sea, after coming to office it undertook a review of the *1979 OCS*. Its official position as stated in the House of Representatives is that "there would be no legal obstacle to the unilateral variation by the Commonwealth Government of the current arrangements under the Offshore Constitutional Settlement".[165] Nevertheless, it considers that the OCS has been working satisfactorily and does not "intend to take action at this time to regain title to the territorial sea. Nor does the

---

164. *Offshore Constitutional Settlement*, *supra*, note 94, at 6, where it is stated, "This grant of proprietary rights and title will ... provide an assurance to the States that the arrangements relating to the territorial sea will have permanency and stability."

165. Australia, House of Representatives, *Hansard*, 20 March 1986, at 1781. The statement was made on March 20, 1986, by Mr. Barry Jones, representing the Minister for Resources and Energy, in response to an Opposition question.

Government intend to alter the current powers legislation".[166]

Notwithstanding this conciliatory approach, the Commonwealth government has made it plain that it is carrying a big stick:

> This approach will remain dependent on the continued satisfactory operation of the existing arrangements which will be kept under review. The question of resuming title may be reconsidered by the Commonwealth should any of the States or the Northern Territory not be prepared to continue acting in ways compatible with the national interest while exercising powers obtained under the OCS. The Prime Minister has recently conveyed this decision to the State Premiers and the Northern Territory Chief Minister.[167]

State officials seem only too aware of the Commonwealth viewpoint, which may help to explain their relatively meek acceptance of Commonwealth amendments to the PSLA.

## Consistency in Laws, Regulations, and Policies

Constitutional uncertainty concerning offshore jurisdiction in 1967 caused Australia's leaders to choose the device of mirror legislation to regulate offshore petroleum. In its review of the regime, the Senate Select Committee noted that the uniformity created by the existence of a "Common Mining Code" was desirable:

> having the obvious advantage that throughout the Australian off-shore waters common principles and procedures, common forms of tenement and common methods of over-all control apply.[168]

Overall, the Committee agreed with the extent of uniformity achievable under the joint legislative scheme and noted that, while the system somewhat reduced the ability of individual states to design their own incentive-oriented exploration schemes, broad discretionary powers provided the DA with a degree of local administrative independence.[169]

Section 23 of the *1967 Agreement*, which required the

---

166. *Id.*
167. *Id.*
168. *Senate Select Committee Report*, *supra*, note 75, at 238.
169. *Id.*, at 246.

Commonwealth and the states to "confer from time to time concerning the operation and administration of the legislation", also reflected the fact that administrative uniformity was seen as a goal; the required consultations seem to have taken place through the Australian Minerals and Energy Council and its Advisory Committee.[170]

The *1979 OCS* is relatively silent on the issues of uniformity and consultation, simply stating in regard to petroleum arrangements within the territorial sea that "the common mining code will be retained as far as practicable".[171] The sections that follow consider the mechanisms for intergovernmental consultation that are utilized under the *1979 OCS* and the extent to which uniformity in petroleum laws and administration, and in other laws, has been achieved throughout the Australian offshore.

### *Mechanisms for Consultation*

Many matters relating to offshore petroleum legislation and administration are presently dealt with through the Australian Minerals and Energy Council [hereinafter AMEC] and its Standing Committee on Offshore Petroleum Legislation [hereinafter AMEC Standing Committee].

AMEC, which pre-dated the *1967 Agreement*, consists of the relevant state and Commonwealth ministers. While it meets approximately every six months, offshore petroleum issues represent only a fraction of its work. Typically, heads of the relevant departments meet in advance of the AMEC meeting to discuss the issues and set the agenda for the Commonwealth and state ministers' meeting.

The AMEC Standing Committee, comprising the heads of the relevant petroleum departments, also meets about every six months. Although its agenda is set by Commonwealth officials, a draft agenda is usually circulated in advance of meetings so that

170. *Id.*, at 156.
171. *Offshore Constitutional Settlement*, *supra*, note 94, at 8.

state officials can suggest additional items. Matters arising at Committee meetings are generally resolved by discussion. Occasionally they are resolved by vote, but the parties recognize that, in relation to Commonwealth adjacent area issues, the Commonwealth effectively has a veto.

When the Commonwealth proposes to amend its offshore petroleum legislation, state viewpoints are generally sought in one of two ways: proposed changes may be circulated to individual states for their comments or the matter may be put on the AMEC Standing Committee's agenda. In the former situation, there is some perception at the state level that too little time is allowed for the preparation of a thoughtful response. This procedure also makes it difficult for the states to consult with one another or to attempt to coordinate their viewpoints. When matters are handled through the Committee, there is greater likelihood of the states arriving at a common opinion, although the fact that Western Australia and Victoria have had the main interest in offshore petroleum issues has sometimes made it difficult to procure state consensus even at the Committee level.

Government officials at both levels view AMEC and its Standing Committee as serving a very important function in the workability of the offshore petroleum regime. These mechanisms provide a regular forum for discussing and resolving problems and, especially in the case of the Committee, have proven essential in working toward consistency in the treatment of all offshore areas.

### *Uniformity in Offshore Petroleum Laws and Administration*

#### *Uniformity Between the Commonwealth Adjacent Areas*

Since all Commonwealth adjacent areas are subject to the Commonwealth PSLA and the Commonwealth is free to amend its legislation as it sees fit, theoretically uniform treatment of all adjacent areas should be easy to attain. Other elements of the Australian regime also contribute to this uniformity.

The *1979 OCS* and the PSLA envisage that day-to-day

petroleum administration will be carried out by the states, largely through discretionary powers vested in the DA. To the extent that the DA's discretion is overseen by the JA, the Commonwealth can influence policies and practices within each adjacent area and thereby promote uniformity when it is deemed desirable. It has been seen that discretion that was formerly lodged in the DA is increasingly being transferred to the JA, enhancing the Commonwealth's ability to pursue consistent policies within the adjacent areas.

In addition, uniformity has been achieved through the use of Administrative Guidelines [hereinafter Guidelines]. Issued by the Commonwealth after some input from the states, these Guidelines cover a broad range of matters including the previously-discussed JA procedures and procedures for the collection and verification of royalties and fees. Among the most notable Guidelines for present purposes are:

- *Exploration Permits — Model Clauses for Work Requirement Conditions*
- *Award of Offshore Petroleum Exploration Permits Under the Cash Bidding System*
- *Criteria for Assessment of Applications for the Award of Petroleum Exploration Permits Under the Work Program Bidding System*
- *Work Program Permit Conditions and Guidelines for Handling Work Program Variations, Permit Cancellations and Permit Surrenders*
- *Cancellation and Refusals to Grant Renewals of Retention Leases*
- *Guidelines for the Declaration of Locations and Related Matters*
- *Guidelines for Release of Data and Information Under Section 118 of the PSLA*
- *Guidelines for Submission of Applications for Approval and Registration of Transfers and Dealings Relating to Titles.*

Officials suggest that the development of the Guidelines was partly motivated by the existence of freedom of information legislation which, it was felt, required a degree of transparency in administrative decision-making. It was also hoped that publicly-available Guidelines would minimize the seeking of judicial review of decisions made pursuant to the PSLA. As well, the Guidelines were intended to promote uniform practices within the Commonwealth adjacent areas. While some of the Guidelines summarize relevant statutory provisions, others set out how administrative discretion is to be exercised in certain situations. For example, Guidelines for variations in work programs, specifying time limits for variation applications and the conditions that must be met in order for variations to be granted,[172] help to ensure that discretion will be exercised in a similar fashion in all adjacent areas. Industry representatives generally applaud the existence of the Guidelines, because they tend to make government decision-making in the Commonwealth adjacent areas more predictable.

The use of "directions" is also of interest in the context of uniformity. Directions were referred to earlier in relation to fiscal matters,[173] but a brief explanation of their place in the offshore petroleum regime will provide a context for the role they play in regard to the issue of uniformity.

The PSLA provides for detailed second-level regulation through two devices, regulations and directions. Section 157 grants the Governor General broad regulation-making power, but very limited use of this power has been made so far.[174] Various explanations are offered for this. One is that petroleum regulations are necessarily very technical and thus time-consuming to draft. It

172. Australia, Dept. of Resources and Energy, *Work Program Permit Conditions and Guidelines for Handling Work Program Variations, Permit Cancellation and Permit Surrenders, 86/4.*
173. See note 140 and accompanying text.
174. At present, regulations only relate to the form of an instrument of transfer and applications for approval of a dealing in title. *See Petroleum (Submerged Lands) Regulations*, Statutory Rules 1985, No. 174 as am. by 1987, No. 293.

is admitted, nevertheless, that highly technical regulations have been drafted pursuant to other Australian statutes. Another explanation relates to the requirement of Australian law that regulations be laid before both the House of Representatives and the Senate. While regulations can become effective before being presented to Parliament, they can be disallowed by either body. Given the long-standing controversy that has surrounded the offshore, some express concern that the presentation of regulations to Parliament might reopen the entire issue of offshore petroleum arrangements. This explanation also seems unconvincing, given that the PSLA itself has been presented for amendment on several occasions since the *1979 OCS*.

A more compelling reason for the failure to develop regulations may be that, under the *1967 Agreement*, regulations had to be agreed to by all governments. Since consensus on regulations proved impossible to achieve, the more informal device of "directions" came into wide use at the state level. While such directions were not necessarily uniform from state to state, they seem to have influenced the present method of detailed regulation in the adjacent areas.

All interests in the Commonwealth adjacent areas have been made subject to the 1985 Uniform Directions[175] that cover everything from safety to drilling and production, the very topics one might expect to be covered by regulations. The Directions, finalized in 1985 through the AMEC Standing Committee, were applied to the various adjacent areas by virtue of s.101(1) of the PSLA which authorizes the DA to give a written direction "as to any matter with respect to which regulations may be made". Many of the provisions in the Directions, an amalgamation of previous directions used by the states, especially by Victoria and Western Australia, are couched in very general language.

---

175. The Directions are entitled *Specific Requirements as to Offshore Petroleum Exploration and Production - 1985*, being a Schedule of the *Petroleum (Submerged Lands) Act* (Cth) [hereinafter 1985 Uniform Directions].

Notwithstanding the application of the 1985 Uniform Directions to all Commonwealth adjacent areas, there remains considerable scope for local variation. The Directions contain reference throughout to the "Director"[176] who, typically, is the state official responsible for the petroleum department. He is the recipient of various reports and has broad authority and discretion under the Directions.[177] Moreover, the Directions are riddled throughout with the term "unless otherwise approved".[178] Although it is not specified by whom such other approval is to be given, in practice it is typically the Director. As a result, individual operators are frequently subjected to a subsidiary type of directions given by the Director in the form of a letter. Not necessarily public, such "letter-directions" often vary or provide additional detail to the Directions. Although s.101(3) of the PSLA prohibits a DA from giving a direction of "a standing or permanent nature except with the approval of the Joint Authority", the rather loose nature of the Directions makes it possible to circumvent or bend this prohibition.

The generality of the 1985 Uniform Directions and their ability to be molded to particular circumstances is seen in some quarters as providing desirable flexibility and facilitating variations required by different operating conditions. At the same time, their general language can lead to differing interpretation and application from one adjacent area to another. For example, it is said that differing state interpretations of the diving directions can cause problems for industry. If such discrepancies lead to difficulties, state officials attempt to resolve them through bilateral

176. Defined in s.103(1) of the PSLA.
177. For example, pursuant to s.283(a) of the PSLA, he is to receive a written report of an event which is not in the normal or ordinary course of a particular operation. Section 203 empowers him to approve updates to safety and emergency response manuals, while s.245 authorizes him to require information at any time concerning the competence of personnel.
178. For examples, see s.502 regarding conformity of equipment to stated standards, s.732 regarding standards to be applied to bridge cranes, s.813(2) regarding manning levels for diving operations, and s.505(2) regarding blow-out prevention.

discussions, failing which the problems may find their way onto the agenda of an AMEC Standing Committee meeting. The Committee thus provides an important means for resolving differing interpretations of the Directions that are problematic.

The direction system, however, can give rise to communication problems between Commonwealth and state officials. Industry members approach state officials when they perceive a need to obtain relief from a requirement in the 1985 Uniform Directions. State officials do not necessarily inform Commonwealth officials that such relief is being granted and, if Commonwealth officials subsequently discover the development of practices of which they were unaware, conflict can result.

It is not clear that the use of the 1985 Uniform Directions as a second level of regulation has been entirely satisfactory. The generality of their standards and the vagueness of their language makes them difficult to enforce and can lead to diversity in application and interpretation. Still, if serious problems occur, the Commonwealth always has the option of passing regulations. This step has not yet been considered necessary.

### *Uniformity Between the Commonwealth PSLA and State Offshore Petroleum Laws*

Consistent with the notion of retaining a "Common Mining Code" for the entire offshore, the major amendments to the PSLA that accompanied the *1979 OCS* were eventually adopted by state Parliaments through passage or amendment of state-PSLAs.[179] As has been seen, the "Common Mining Code" is no longer required for the reasons of legal uncertainty that necessitated it under the *1967 Agreement*. Rather, it is intended to provide uniform regulation of offshore petroleum operations wherever they take place. To a considerable extent, the PSLA and its state

179. The counterpart Victorian and Western Australian laws are the *Petroleum (Submerged Lands) Act 1982* (Vict.) (No.9772) and the *Petroleum (Submerged Lands) Act 1982* (W.A.) (No.33).

counterparts have remained similar. Some key divergences have crept into the system, however, two examples of which are discussed here.

In 1985, the Commonwealth legislation was amended to permit cash bidding in highly prospective offshore areas.[180] This system was controversial: some state governments felt that cash bidding, in lieu of work programs, would discriminate against smaller petroleum companies and detract from the amount of exploration money available. With this differing exploration philosophy, it is not surprising that the states have declined to make provision for cash bidding in the petroleum legislation that applies to their coastal waters. Commonwealth officials see this divergence between the state and Commonwealth statutes as insignificant, since cash bidding is optional and is not expected to be used heavily.[181]

The other major variation has resulted from the Commonwealth's introduction of the RRT, discussed earlier.[182] Here the difference between the state and Commonwealth governments is less due to philosophy than to uncertainty about how the proceeds from the RRT will be shared. Although the RRT was introduced through legislation other than the PSLA, it necessitated amendments to the PSLA to deal with those situations where the *Royalty Act* will not apply.[183] These provisions, of course, have no counterpart in the state offshore petroleum statutes.

---

180. The cash bidding amendments are contained in s.22A to 22C of the Commonwealth PSLA. For a discussion of these provisions, *see* J.C. Starkey, "Australia's Offshore Petroleum Legal and Administrative Regime" (1987) 27(2) A.P.E.A.J. 25 and P.C. Reid, "Recent Changes to Australia's Offshore Petroleum Regime" (1986) 26(1) A.P.E.A.J. 102 at 103.
181. The utilization of cash bidding has given rise to another issue, namely whether the cash bidding proceeds belong to the Commonwealth or the adjacent state. In support of their entitlement, the states rely upon the *1967 Agreement* which provided, in clause 22, that the states would be entitled to "all moneys, other than royalties, payable under the 'Common Mining Code' in relation to the adjacent area of the State".
182. See text accompanying note 142, *supra*.
183. For example, ss.39A and 40A were added to the Commonwealth PSLA.

These examples demonstrate that, in a relatively short period, the two levels of government have developed different views on how to encourage exploration and what fiscal instruments are appropriate; so far, it has not proven possible to reconcile their points of view. Legally, the Commonwealth changes have been easy to handle since amendment of the PSLA is clearly within Commonwealth jurisdiction. Apparently, these are situations where, under the terms of *1979 OCS*, it has not been "practicable" to retain a "Common Mining Code" for the Commonwealth adjacent areas and the state coastal waters.

The goal of a "Common Mining Code" can be undermined in other ways, such as the length of time it can take a state Parliament to mirror changes that have been made at the federal level. For example, despite the many amendments of a technical nature that have been made to the Commonwealth PSLA in the past few years, Western Australia's PSLA-counterpart law has not been amended since 1982, while the Victorian Parliament took several years to bring its PSLA-counterpart legislation into line with the Commonwealth PSLA. Such delays are not the result of disagreement about the desirability of such technical amendments, but reflect the reality that different governments have different legislative programs and priorities. It is not beyond imagination that the amending process will begin to occur in reverse, with states taking the initiative in altering their offshore petroleum laws and suggesting that the Commonwealth and other states follow suit. If this begins to happen in a significant way, the concept of a "Common Mining Code" may become increasingly fractured.

Beyond the matter of statutory commonality lies the question of administrative commonality. Some states (for example, Western Australia) follow the practice of applying the 1985 Uniform Directions even to offshore areas within their jurisdiction. Moreover, the major role played by the state mining departments and by the state minister (in his capacity as decision-maker under the state-PSLA and as DA under the Commonwealth PSLA) has contributed to uniformity in practice. As the authority of the JA increases, however, this may be less the case. While the states will be aware of how discretion is exercised by the JA in the Commonwealth adjacent area, they will not necessarily wish to follow the same path, particularly if the state view has not carried the day in the JA.

Divergence in the law and practice within the two areas off the same state has been mostly an academic problem offshore Victoria, since the bulk of offshore activity has occurred in the adjacent areas subject to the Commonwealth PSLA. This situation is changing, however, as prospective blocks spanning the coastal waters and the adjacent area are released for exploration. As discussed earlier, there has been considerably more activity in Western Australia's coastal and internal waters, and some of the practical and legal problems that can accompany the loss of a "Common Mining Code" have been experienced there already.

*Uniformity Between State Coastal Waters*

In theory, these areas too are governed by a "Common Mining Code" and, since the states participate in the administration of the Commonwealth adjacent areas, there should be an impetus toward a common approach. At the least, there should be considerable knowledge about the practices followed in other states.

Just as there may begin to be divergence between the Commonwealth and state laws, however, there may be divergence between the state offshore petroleum statutes. The existence of AMEC and its Standing Committee should continue to provide an important forum for the exchange of view and ideas.

*Summary*

Prior to the *1975 Seas and Submerged Lands Case*, uncertainty about which level of government had offshore jurisdiction made it legally important to have identical Commonwealth and state offshore petroleum laws. Under the *1979 OCS*, uniform petroleum laws throughout the offshore areas remain an important goal for reasons of administrative simplicity. All adjacent areas are covered by the Commonwealth PSLA and the existence of Guidelines and the 1985 Uniform Directions helps to ensure that similar administrative practices are followed in each adjacent area. Discretion vested in the DA provides some scope for variation, and differing interpretations and approaches can sometimes cause problems for industry. AMEC and its Standing Committee have provided a mechanism for discussing serious problems.

Despite the role of AMEC and its Standing Committee, the states have refused to follow the Commonwealth lead by applying cash bidding and the RRT to their coastal waters. Moreover, the states have not been quick to reflect technical amendments to the PSLA in their offshore petroleum laws. With the growing power of the JA, some states may begin to pursue their own policies in the coastal waters, particularly if their views are overruled in the JA. The result could be divergences between the policies pursued in the various offshore areas.

These gradual differences can cause problems for industry, particularly where prospective areas overlap boundaries. It is too soon to tell, however, how difficult it will be to maintain a "Common Mining Code" for the Australian offshore.

***Uniformity In Other Laws***

There has already been reference to various provisions that pertain to application of non-petroleum laws in the offshore areas. To reiterate briefly:

1. Sections of the *State Powers Act* give the states their normal legislative authority within the coastal waters, as well as more limited powers (for example, in relation to certain shipping and fisheries matters) in the area beyond.
2. Section 9 of the PSLA applies state law in adjacent areas in regard to matters connected to petroleum activities. Exceptions to this include the PSLA itself, laws inconsistent with Commonwealth laws, and taxation laws. Commonwealth

regulations can also override the application of state laws.

As a result of these provisions, it may be concluded that the non-petroleum laws that apply offshore will vary as much as the laws of one state differ from those of its neighbour. To take one example, a state's workers' compensation laws apply to its Commonwealth adjacent area and its state coastal waters, but such laws will not necessarily be the same as those that apply offshore a neighbouring state. While this has occasionally caused problems for industry (for example, when a nurse's qualifications were acceptable off one state but not another), for the most part such variations do not seem to have been considered troublesome.

Problems that might arise in regard to vessels that move from one offshore area to another have been largely, although not entirely, resolved under the Australian system. Because of the Commonwealth's general jurisdiction in regard to interstate trade and commerce, and provisions in the *1979 OCS*,[184] the Commonwealth has regulatory powers in relation to marine aspects of the offshore with the exception of those vessels that only operate in one adjacent area. Thus, generally uniform standards apply.

## Intergovernmental Relations

Earlier sections have discussed the extent to which there is interaction between the state and Commonwealth governments in the context of JA decision-making and the operation of AMEC and its Standing Committee. For the most part, these interactions appear to occur smoothly. Nevertheless, areas of friction emerge from time to time.

The Senate Select Committee reported that while the level of cooperation under the *1967 Agreement* had been quite good, the Commonwealth government at times experienced delays in receiving information from state authorities. The Auditor-General has pointed to problems of information flow from the state to

---

184. See discussion *supra* in the text following note 98.

Commonwealth government in relation to fiscal matters.[185] Over time, some of the information flow problems have been addressed through strengthening the Commonwealth's entitlement to information in the PSLA itself. Problems have sometimes developed as well in regard to the interaction of one level of government with the industry without the knowledge of the other. Examples have been given earlier of this phenomenon in the context of letter-directions to operators from state authorities and one finds similar concerns at the state level regarding Commonwealth dealings with the industry on the crude oil levy. It is not obvious what the solution to such problems is, but at least there exists an established mechanism through which such matters can be discussed and, hopefully, resolved.

In the area of environmental impact assessment, a cooperative approach has developed between the two levels of government.[186] Under the Commonwealth *Environment Protection (Impact of Proposals) Act 1974*,[187] the Minister is responsible for ensuring that matters affecting the environment to a significant extent are fully examined and taken into account. In recognition of this responsibility, drilling applications are referred to the Commonwealth for determination as to their environmental significance. When resource projects are involved, the Commonwealth relies upon state environment authorities for advice.

From the Commonwealth viewpoint, the results have been satisfactory:

> The Department considers that present consultative arrangements at Federal, State and industry levels for the environmental assessment of resource projects have been effective in achieving environmental objectives in an efficient and cost-effective manner. Environment protection issues have been fully considered and the potential for

---

185. *Auditor-General's Report*, *supra*, note 110, at 15 *ff*.
186. The information which follows is derived from a Commonwealth document entitled "Administrative Arrangements for Assessing the Environmental Impact of Petroleum Exploration and Development in Australia" [unpublished].
187. *Environment Protection (Impact of Proposals) Act 1974* (Cth) (No.164).

duplication and undue costs and delays has been kept to a minimum.[188]

Satisfactory cooperation has also been experienced in relation to potential prosecutions under the PSLA. Although there have not yet been any prosecutions, there has been experience in carrying out investigations with a view to prosecuting. The state inspectors are appointed under the PSLA as Commonwealth agents, and, given its day-to-day administrative role, the state department's heavy involvement is required in pursuing possible breaches of the Commonwealth PSLA. With the consent of the Commonwealth Department, the state department consults the Commonwealth Solicitor for advice. If Commonwealth approval is obtained, a prosecution could then be initiated, with the Commonwealth Solicitor acting on the state's instructions.

### Overlap or Lack of Clarity in Legislative Authority

For the most part, the Australian system seems to have dealt successfully with the question of which government's laws apply to the offshore. Clear language in the Commonwealth PSLA applies state law in the Commonwealth adjacent areas in regard to matters connected to petroleum activities and the *State Powers Act* gives states legislative authority within the state coastal areas. Thus, while the carving up of the Australian offshore into a myriad of areas has caused some problems, the legislators have managed to express unambiguously their intentions as to what laws will apply in each.

### Review of Decisions

Because of differences between the state and Commonwealth offshore petroleum laws, this section does not purport to provide a comprehensive examination of the way in which decisions can be reviewed under all the relevant statutes. Instead, the emphasis here is on the provisions of the Commonwealth PSLA and other

---

188. *Id.*, at 3.

Commonwealth laws. The PSLA contemplates the review of certain decisions by the Commonwealth minister, the courts, or the Administrative Appeals Tribunal.

***Review by Commonwealth Minister Pursuant to Section 118***

Section 118(5A) of the PSLA authorizes the DA or the Commonwealth minister to make publicly known or available upon request to a person, after five years, information in a furnished document:

> being information that relates to the sea-bed or subsoil, or to petroleum, in a block, and that, in the opinion of the Designated Authority or the Commonwealth minister, is a conclusion drawn, in whole or in part, from, or an opinion based, in whole or in part, on any information contained in a document ....

Section 118(5B) prescribes a procedure to be followed before the DA or Commonwealth minister may release such information, which procedure includes the opportunity for an objection to be filed by an interested person. In relation to a decision by a DA concerning such an objection, s.118(5G) gives the Commonwealth minister a review power upon the request of a person dissatisfied with the DA's decision. The decision must be reviewed within forty-five days, and the Commonwealth minister may revoke the DA's decision or substitute his own. Pursuant to s.118(5L), he must notify the objector of the result of his review, giving written reasons. These provisions have yet to be utilized.

***The Role of the Courts***

The PSLA gives the courts original jurisdiction to determine certain issues (such as the proportion of a petroleum pool to be allocated where a pool crosses licence areas (s.6)) and other matters such as the extent of compensation payable to a licensee who is directed to make changes to its pipeline (s.72). Courts also have appeal powers pursuant to s.88 in relation to registration matters.

At a more general level, the courts have powers of judicial review pursuant to the *Administrative Decisions (Judicial Review)*

*Act 1977*,[189] which applies to all Commonwealth legislation. Although s.151 of the PSLA vests state courts with federal jurisdiction "with respect to matters arising under this Act or the regulations", s.9 of the former statute removes state court jurisdiction "notwithstanding anything contained in any [other] Acts". Thus, s.151 of the PSLA probably has to be read down to accommodate this provision, leaving review powers pertaining to the PSLA in the federal court.

***Review by the Administrative Appeals Tribunal***

Since 1975, the Commonwealth *Administrative Appeals Tribunal Act*[190] has made it possible for decisions of federal bodies or individuals to be subjected to substantive review by the Administrative Appeals Tribunal [hereinafter AAT]. These substantive review provisions have been made applicable to decisions taken by the Commonwealth minister exercising the powers of a DA in relation to certain Commonwealth Territories, exercising certain powers relating to the release of information under s.118, and exercising certain review powers under s.152.

What is of particular interest is the limited applicability of the AAT procedure under the Commonwealth PSLA. At the time these amendments were being considered, two (apparently) contradictory points of view were put to the Commonwealth minister.[191] On the one hand, it was proposed that no AAT review should be taken, since protection was already available under the *Administrative Decisions (Judicial Review) Act*. The minister rejected this view, because successive governments have taken the decision "that administrative decisions should be reviewable by the AAT on the merits of individual cases".[192] Another opinion was that JA and DA decisions should also be subject to the AAT procedures. The minister responded:

---

189. *Administrative Decisions (Judicial Review) Act 1977* (Cth) (No.59).
190. *Administrative Appeals Tribunal Act 1975* (Cth) (No.91).
191. Australia Mining and Petroleum Law Association Ltd., 4(3) *Bulletin* 37.
192. *Id.*

> This is not acceptable as it would make decisions by State and Northern Territory Ministers reviewable under legislation intended only to apply to decisions made by Commonwealth Ministers and delegates of those Ministers. Further, if Joint Authority decisions were reviewable, major policy decisions would become reviewable by the AAT.[193]

As has subsequently been observed, the latter justification somewhat begs the question, since the very purpose of the AAT legislation is to provide a substantive review of the decisions subjected to it.

---

193. *Id.*

# 4. THE CANADIAN AND AUSTRALIAN OFFSHORE PETROLEUM REGIMES COMPARED

Canada and Australia have resolved intergovernmental disputes concerning the management of offshore petroleum in quite different ways. This section begins by contrasting some of the geographical, political, historical, and legal forces that have influenced the evolution of each country's system.

Notwithstanding these differences, each regime has given rise to remarkably similar problems, many of which have important implications for industry. The remainder of the section considers a series of issues arising under both systems and examines the approach adopted in the two countries. The topics dealt with are the forms taken by the political arrangements and what implications these have had for the resolution of later intergovernmental disputes; the constitutional status of the political arrangements; the way in which legislation, regulations, and other instruments can be amended under the two systems and what effect this may have on the flexibility of the management regimes; the extent to which the two systems have facilitated the achievement of consistent laws, regulations, and policies between the various offshore areas; the decision-making and review mechanisms; the resulting relationships; the clarity of legislative jurisdiction under the two regimes; the fiscal relationships under each; and the boundary issues that have arisen and how they can be resolved. Underlying the analysis throughout is the question of what impact these matters have had on industry, and with a view to the concluding section of the paper, what lessons each country might take from the experiences of the other.

**Geographical, Historical, Legal, and Political Differences**

Australia's system is multilateral, with all six states and the federal Northern Territory having participated in the *1979 OCS*. This may be largely a result of the geographical fact that all these jurisdictions have coastlines. Although as of 1989 not all jurisdictions had a major interest in offshore petroleum resources (as has been seen, Victoria and Western Australia have had the greatest interest so far), the mid-1960's negotiations led to an offshore regime that embraced many matters in addition to petroleum. Consequently, all the states had a strong interest in the negotiated arrangements. The fact that the *1967 Agreement* was negotiated on a multilateral basis seems to have set the stage for what followed: the states jointly attacked the *Seas and Submerged Lands Act* in litigation, and, following the *Seas and Submerged Lands Case* decision, it was nearly inevitable that all the parties would participate in the accommodation that was reached through the *1979 OCS*.

In Canada, there were early efforts to resolve the constitutional issues on a multilateral basis, at least as regards the east coast. The *1964 Agreement* on offshore boundaries reached between the premiers of five provinces suggests that, at one time, a multilateral solution might have been possible. A multilateral approach was also pursued in the discussions that resulted in the ill-fated *1977 Understanding* between Canada, Nova Scotia, New Brunswick, and Prince Edward Island. At this point, however, two factors seem to have destroyed the future possibility of a multilateral solution. One was the focus upon petroleum resources which, given the available information, made the arrangements of much greater significance to some provinces than to others. The other was the different provincial constitutional and legal histories. Newfoundland, in particular, was so convinced about the strength of its legal position that it was unwilling to accept arrangements that were satisfactory to the other three provinces. Once the *1982 Nova Scotia Agreement* had been negotiated between Canada and Nova Scotia, a bilateral approach was hard to avoid. Moreover, the relatively junior status of the territorial governments made it unlikely that, at that time, the federal government would have considered granting them the sort of rights it was prepared to share with the provinces.

Differing legal and political dynamics have influenced the outcome in the two countries. The 1975 *Seas and Submerged Lands Case* decision made the legal parameters of offshore jurisdiction quite clear in Australia; that a Commonwealth government, in 1973, turned its back to an intergovernmental arrangement agreed to by an earlier government and enacted the *Seas and Submerged Lands Act 1973* must be seen, at least in part, as a reflection of a widely-held perception that the previous political solution was not in best interests of the country. In Canada, the legal position became clearer after the *Hibernia Reference*, but there remain some legal uncertainties, especially as regards the territorial sea. Those uncertainties, as well as the relatively strong political position occupied by Canadian provinces, have resulted in intergovernmental arrangements for offshore petroleum management that, to a considerable degree, side-step what the Supreme Court has said about the legal rights of the two levels of government.

## Issues Arising Under Both Systems

### The Form Taken by the Political Agreements

Although the substance of the political agreements is clearly more important, overall, than the form they have taken, the differing forms adopted have not been without significance.

The *1979 OCS* in Australia is evidenced by a very informal document that merely sets out agreed principles. Many of these principles are extremely vague, for example, the provision that simply adopts the royalty arrangements of the *1967 Agreement*. It is arguable that this casual approach has already worked to the detriment of the states. Recall, for a moment, the litigation between the State of Victoria and the Commonwealth concerning the JA's power to instruct the DA on methods of royalty calculation.[194] Among other things, the state attempted to argue that the JA's exercise of its power contravened the "sharing" notion referred to in a preamble to the PSLA. Had the *1979 OCS* been evidenced in a more detailed and formal document, it is possible that the state might have had more success with this argument than it did. Another example concerns the Commonwealth-state disputes over revenue-sharing of the RRT and the proceeds of cash bidding. The vagueness of statements concerning revenue-sharing in the Agreed Arrangements[195] means that the entire matter is now up for renegotiation only a decade after the *1979 OCS* was devised. A similar point can be made about the gradual diminution of the DA's role and the corresponding enlargement of that of the JA: since the Agreed Arrangements did not specify exactly which body would have what powers, except in the most general terms, state arguments that the Commonwealth is contravening the *1979 OCS* have only a narrow foothold. Viewed one way, the vague language of the *1979 OCS* permits a lot of flexibility to adjust the arrangements as experience dictates. So far, however, this has seemed to work in the interests of the party with the greatest bargaining power, namely, the Commonwealth.

In contrast, the Canadian Accords provide great detail about the intended arrangements, for example, listing the functions that are to be vested in the Offshore Petroleum Boards and specifying what provincial laws are to be incorporated by reference in federal legislation. The intentions of the parties to the Accords being relatively clear, any government wishing to unilaterally amend its enabling legislation would face very difficult negotiations unless it could convince the other that the proposed amendments would be in the interests of both. In the event of litigation, the uncertain status of intergovernmental agreements in Canadian law makes it far from clear that a protesting province's legal arguments would succeed (especially as regards the continental shelf); but the

---

194. *Fordham* v. *Evans*, *supra*, notes 105 and 124.
195. *Supra*, note 94.

detailed language of the Accords would provide Newfoundland or Nova Scotia with more political and moral ammunition than the Australian states could likely muster in a similar situation.

### The Constitutional Status of the Intergovernmental Arrangements

The Constitution of Australia has never been amended formally to reflect the terms of the *1979 OCS*. Nevertheless, the OCS has been enacted in such a way that there exists an argument that the Commonwealth cannot amend the arrangement unilaterally (namely, because the *State Powers Act* was passed as a result of "request" legislation from the states). Beyond this legal argument lies the possibly more important fact that, in relation to the states' title to the coastal waters, compensation would likely be required should such title be diminished. It should be borne in mind that the *1979 OCS* regime in Australia came into being largely because a subsequent Commonwealth government had unilaterally passed legislation that contradicted the *1967 Agreement*, an Agreement lacking such quasi-constitutional protections.[196]

From another vantage point, the Commonwealth is in an unassailable legal position so far as the continental shelf is concerned and seems prepared to rely upon this fact to pursue new

---

196. Various explanations can, of course, be offered in defense of the Commonwealth's decision to pass the *Seas and Submerged Lands Act 1973*. First, the *1967 Agreement* had preserved the legal position of all the parties. Second, in its 1971 *Report*, the Senate Select Committee had supported the desirability of seeking a legal resolution of the offshore issue. Third, by then the *B.C. Offshore Reference* had been decided, lending support to the notion that the Commonwealth's legal position was stronger than that of the state. Fourth, the states had attempted themselves to have the offshore issue resolved by the Privy Council even before the *Seas and Submerged Lands Act 1973* was introduced. Notwithstanding these justifications, the *1967 Agreement* had contained a commitment by all parties to engage in continuing consultations concerning the Agreements and to alter the mirror statutes only by agreement. The conclusion that the passage of the *Seas and Submerged Lands Act 1973* by the Commonwealth contravened the very basis of the *1967 Agreement* is, therefore, inescapable.

policy initiatives and amendments to the PSLA. So far, this has worked well enough: the Commonwealth has not interfered with the territorial sea elements of the *1979 OCS* and the states have not complained loudly about Commonwealth-initiated changes to the PSLA. Still, it is worth noting that the alterations have been substantial (such as the enlargement of the JA's role, the elimination of Western Australia's special decision-making rights, and the imposition of the RRT and cash bidding despite state opposition).

What is the position in Canada? Until such time as Newfoundland and Nova Scotia take the initiative in seeking constitutional entrenchment of the Accords, they run the long-term risk that changed circumstances could cause a later federal government to renounce them. While such a development would have severe political consequences, the Australian example shows that such events can occur. Indeed, a Canadian case demonstrates the same point.[197] As pointed out earlier, any unilateral federal renouncement of the Accords could exhume the issue of jurisdiction over the territorial sea off Newfoundland, and, *vis-a-vis* Nova Scotia, could lead to litigation concerning the entire offshore, notwithstanding the fact that as of April 1989 Canadian jurisprudence regarding the offshore has favoured the federal position.[198] If constitutional entrenchment of the Canadian

---

197. See *Re Taxation Agreement Between Government of Saskatchewan and Government of Canada* (1945), [1946] 1 W.W.R. 257 (Bd. of Arbitration). The case involved a situation where the federal government declined to comply with a taxing arrangement that had been agreed with the Province of Saskatchewan. It was found by two of the three members of a Board of Arbitration that the province was not entitled to a remedy when federal failure to pay was based upon a set-off in relation to provincial default under another agreement. While the Arbitration Board seems to assume the enforceability of these intergovernmental agreements, it should be noted that they had been entered into pursuant to statutory authorization.
198. As has been explained above (see text accompanying note 11 *ff.*), the Supreme Court of Canada has ruled specifically only on the question of offshore jurisdiction as regards British Columbia and the continental shelf off Newfoundland. Some commentators see these decisions as having much broader import. Harrison, for example, concludes that "the constitutional status of all offshore areas has been settled, with only

Accords were pursued, innumerable practical, legal, and constitutional matters would need to be addressed.[199]

### Amendment of the Enabling Laws and Regulations

Under the scheme developed in Australia, it is (at least in theory) relatively easy to amend the offshore petroleum legislation. Since the regime worked out in the *1979 OCS* is not based on the idea of mirror legislation, as was the *1967 Agreement*, there is no legal reason why the petroleum statutes passed by the two levels of government must remain identical. In relation to the adjacent areas subject to the Commonwealth PSLA, there are no legal constraints on the Commonwealth's freedom of action: the High Court has said these areas are within Commonwealth control and the *1979 OCS* does nothing to alter this result. So far, the Commonwealth has chosen to seek the views of the states on proposed amendments, often through AMEC's Standing Committee; and while the states' submissions have undoubtedly been considered,

---

limited exceptions." He notes four exceptions: the three-mile territorial sea adjacent to Newfoundland; possible pre-Confederation extensions of specific boundaries, such as those established by British Columbia; certain areas where jurisdiction may have to be determined on a case-by-case basis in accordance with the common law of waters *inter fauces terrae*; and other areas such as Sable Island off Nova Scotia. He states, "Apart from the question of Sable Island, however, none of these exceptions to the general assertion that the constitutional question is essentially settled is really relevant to those offshore areas where the major oil and gas activity is being undertaken." See Harrison, *supra*, note 19, at 79. Hogg takes a slightly less extravagant view of the implications of the offshore cases, arguing that

> the Supreme Court's reaffirmation of the rule that provincial boundaries end at the low-water mark, and its denial to Newfoundland of international statehood at the time of union, make it unlikely that the Court will reach any different decision with respect to the territorial sea. As well, the Court will probably want to avoid the anomaly that the territorial sea off the coast of one province has a different constitutional status than the territorial sea off the coast of the other provinces."

Peter W. Hogg, *Constitutional Law of Canada*, 2d ed. (Toronto: Carswell, 1985) at 588.

199. One of these issues (namely, which of the amending formulae would apply) is referred to *supra* in the text accompanying note 22. Other issues include the following: What form would entrenchment of the Accords take? What provision would be made for amending the petroleum and related legislation that implements the Accords?

there is no evidence that the Commonwealth's policy choices have been particularly influenced by state viewpoints. As to second-level regulation, whether in the form of directions or regulations, the Commonwealth is under no legal obligation to consult the states, although, in fact, the states played a major role in the development of the 1985 Uniform Directions.

In relation to the coastal waters, so long as the *State Title* and *State Powers* statutes are in effect and in the absence of conflicting Commonwealth legislation, the states have the legal authority to alter their regimes as they see fit. Because of the notion of a "Common Mining Code" in the *1979 OCS*, they will also be inclined to discuss proposed changes with the Commonwealth and the other states but, at the end of the day, they can pursue their preferred policy paths.

The situation in Canada is much more complicated, both legally and practically. The practical complications are discussed in the next section in relation to the issue of uniformity. From a legal viewpoint, the enabling legislation reflects certain of the Accords' requirements that both levels of government concur in amendments to the Accords, the legislation, and specified kinds of regulations. So long as such statutory requirements exist, the stated process must be followed. In Canada, moreover, "mirror" legislation may be desirable legally in relation to those offshore areas where there has been as yet no specific judicial determination as to jurisdiction; it can be argued that legal security can only be given for areas not covered by the *Hibernia Reference* through legislation passed by both levels of government. As was the case in Australia under the *1967 Agreement*, this suggests a need for the federal-provincial statutes to remain in complete harmony.[200]

---

200. To the extent that one sees the offshore cases as suggesting jurisdictional results even for areas where the issues have not been litigated (*see*, e.g., Harrison and Hogg, *supra*, note 198), the need for the "mirror" type of legislation is less compelling. Still, the Canadian jurisprudence has not, so far, led to such a clear result as the *Seas and Submerged Lands Case* in Australia. A conservative view of the Canadian jurisprudence, therefore, supports the notion that "mirror" legislation plays a useful function in

Provided the two levels of government achieve consensus on the need for technical amendments or the desirability of pursuing particular policy directions,[201] it may be relatively easy to amend the statutes. The experience in Australia with a mirror legislation scheme, however, suggests that there may be long time periods required to achieve intergovernmental agreement on complex matters such as detailed regulations. Additionally, differences in the electoral situations and the legislative priorities and programs of the two levels of government may make it difficult to ensure that, at any given time, the two sets of statutes or regulations are a perfect reflection of each other.

## Consistency in Laws, Regulations, and Policies

Under both regimes, the involvement of two levels of government has resulted in a highly complicated and somewhat fractured system of resource management. In Australia, the entire Commonwealth adjacent area is subject to the Commonwealth PSLA and (to the very limited extent that they exist) one set of regulations. For administrative purposes, however, the Commonwealth adjacent area has been carved up among the states, with state departments having considerable freedom to pursue their own policies in relation to day-to-day matters. While this could lead to wide variations from one adjacent area to another, in

---

regard to at least some of the offshore areas. While it may be (as Harrison suggests, *supra*, note 198) that such areas are not at present of great interest to the petroleum industry, that situation could change.

201. The only significant experience to date with amending the "mirror" statutes relates to the controversy over drilling on Georges Bank, as described in the section entitled "Relationships Between the Offshore Petroleum Boards, the Federal and Provincial Governments, and the Federal and Provincial Agencies". As was pointed out, in that instance the provincial statute was amended prior to its proclamation while the federal Bill was altered at the Committee stage to impose a moratorium. Harrison has drawn attention to the fact that the moratorium originated with a unilateral provincial initiative, leaving "industry to puzzle over the ultimate impact of regional accords on the ability of the federal government to maintain a consistent regime across the frontier lands." *Supra*, note 57, at 14.

practice an unacceptable degree of deviation from a common approach has been prevented by the application of the 1985 Uniform Directions throughout the adjacent area and the development of Commonwealth Guidelines to guide the exercise of discretion. The DA/JA structure of decision-making, moreover, ensures that, in relation to important matters, the Commonwealth minister can exert a unifying or harmonizing force. With the gradual expansion of the role of the JA and the corresponding loss of power of the DA, consistency in administrative practice among the adjacent areas will be enhanced.

Somewhat less success has been achieved in relation to consistency in the petroleum laws that govern the adjacent areas and those that govern the state coastal waters. Despite the existence and satisfactory operation of consultative mechanisms such as AMEC and its Standing Committee, policy differences have developed between the states and the Commonwealth governments on issues such as cash bidding and the RRT. These differences have been tolerated but have caused complications, especially where there are prospective areas that overlap boundaries. At a more routine (but nevertheless significant) level, it has proven impossible to keep the state offshore petroleum laws in perfect tune with the Commonwealth PSLA: many of the changes introduced to the PSLA have gradually filtered down to the state level, but in some cases time delays have been significant. Australia's earlier experience with mirror legislation suggests that there is no easy solution to this problem. As cracks in the "Common Mining Code" begin to appear, it might be expected that differences between the state petroleum laws governing the coastal waters will also increase. To the extent that there is a lack of consensus between the state and Commonwealth members of the JA, moreover, one might expect to find divergent policies and administrative practices developing in the two areas offshore a particular state.

Within the Australian system, state laws in relation to matters other than petroleum exploration generally apply throughout the entire area off an individual state. As between the states, non-petroleum laws differ in the offshore area to the same extent that they do onshore, although this does not appear to have caused serious problems.

Under the Canadian system, the offshore has been carved up in a somewhat different way, but in a way that gives rise to its own problems of consistency. In relation to areas off each of Newfoundland and Nova Scotia, it will be a challenge to retain a perfect mirror image between the relevant federal and provincial statutes at any given moment. As has been seen above, there are legal reasons why this may be desirable under the Canadian system, at least in relation to areas other than the Newfoundland continental shelf. An even greater challenge will be to retain consistency as between these two offshore areas and other offshore areas such as the Arctic and the west coast, particularly in view of the fact that day-to-day administration will be handled by independent regional Offshore Petroleum Boards. Industry may have to bear additional internal costs as a result of minor differences between the two east coast regimes, and, if so, the situation could be exacerbated as bilateral arrangements between the federal government and British Columbia, the Yukon, and the Northwest Territories are brought to fruition. Federal authorities will have a heavy responsibility to minimize unnecessary differences in the regimes, to promote discussion and consultation, and to attempt to achieve consensus on contentious issues.

There is no attempt under the Canadian system to have consistent social laws or fiscal arrangements apply to the two offshore areas; this does not seem to be a major concern for the industry.

### Decision-Making and Review Mechanisms

Under the Australian regime, petroleum management decisions are handled in a relatively straightforward way. Day-to-day matters in the Commonwealth adjacent areas are determined by state authorities through the medium of the DA, while more important issues are resolved at the JA level. Commonwealth Guidelines have helped to specify the way discretion will be exercised in relation to some matters, and the establishment of the AMEC Standing Committee has proven an important way of facilitating communication. Within each state's coastal waters, state authorities have the decision-making power. A high degree of satisfaction seems to exist in relation to decision-making by state authorities; more concern is expressed, both by state and industry officials, about the sometimes slow and cumbersome nature of JA decision-making. It is possible that this type of complaint will escalate as greater power is centralized in the hands of the JA.

Judicial review is available in relation to specified issues (such as pooling and registration), while decisions made under the PSLA are subject to general Commonwealth laws concerning judicial review. On the other hand, the Administrative Appeals Tribunal procedures have been made applicable to PSLA matters in only a limited way.

The decision-making arrangements under the Canadian regime are incredibly complicated. Much of the power over day-to-day matters is vested in the Offshore Petroleum Boards, but, as has been seen, their authority in relation to many matters is subject to external control. Another layer of complexity exists in the Canadian system because the arrangements for the two Boards, while similar, are not identical. Some steps have been taken to open lines of communication between the two provinces, but as a Board has yet to be established in Nova Scotia, it has been difficult to make much progress in this direction. A further problem under the Canadian regime results from the attempt to transplant the Oil and Gas Committee from the OGPCA to the Accord regime where it fits, at best, awkwardly. This situation is partly a result of the time constraints under which the *Atlantic Accord Act* had to be drafted coincided with major policy changes at the federal level, making it impossible to work out the implications of old structures for new administrative arrangements. Ironically, the situation is also a result of the federal government's justifiable desire to keep a common core of legislation for all offshore areas. It remains to be seen how easy it might be to revise the decision-making arrangements, now that statutes at both the federal and provincial levels require tandem alteration.

It is worth recalling that, after only a few years of experience in Australia, the Commonwealth government has deemed it desirable to streamline and simplify aspects of its offshore petroleum decision-making mechanisms even though, in comparison to Canada's, they were already quite straightforward. To recapitulate, Western Australia's special rights have been obliterated and a list of matters that could lie with the JA or the DA have now been clearly assigned to the JA.[202] In regard to the latter changes, the Commonwealth minister's rationale was that industry required a clearer system of decision-making, while in regard to the former, it was felt that all states should be treated the same. One must wonder whether, over time, the Canadian system (which is much more complex to start with) will also require simplification. One must also ask whether some of the minor differences between the arrangements for Nova Scotia and Newfoundland that do not serve a particularly important function (as revealed by a comparison between Tables A and C) should be reconsidered. Furthermore, Australia's experience in this regard should be kept in mind by federal officials who are negotiating the arrangements with the British Columbia and territorial governments, so that insignificant differences between the regimes can be kept to a minimum.

Industry's experience in Australia suggests that the involvement of two levels of government in the offshore necessitates sensitivity to intergovernmental relations and, often, a commitment of time and resources to ensure that industry's interests are not lost in the face of intergovernmental disputes or power struggles. Undoubtedly, the same situation applies in Canada.

---

202. See discussion in the section entitled "Decision-Making Mechanisms".

Relationships

Under the Australian system, only government departments and ministers are involved in decision-making. The multilateral nature of the system means that there are many such institutions with a role to play but, since only a few states have a strong interest in petroleum matters, in reality most of the interaction on key issues has occurred between the Commonwealth and two or three state governments. All those involved stress the critical role that has been played by intergovernmental consultative mechanisms such as AMEC and its Standing Committee. Not only have these bodies provided an obvious forum in which the states can put forward their views and the Commonwealth can test its developing ideas; perhaps even more importantly, the personal contacts nurtured by these institutions have made it easier to deal with the interstate problems that occasionally arise. The Commonwealth-state relationships in regard to offshore petroleum have not been uniformly smooth since the *1979 OCS*, especially when the Commonwealth has taken initiatives that do not find favour with the states. It seems to be a widely-held view that these unpopular actions on the Commonwealth side have been accepted by the states without major political battles because the legal authority of each level of government is clear. The *Seas and Submerged Lands Case* decision, therefore, is considered to have reduced some of the tensions between the two levels of government.

Under the Canadian system, federal and provincial departments and ministers also play an important role in management of offshore petroleum.[203] The Memoranda of Understanding that have been provided for in the Canadian legislation should help to reduce interdepartmental rivalry and promote a one-window approach to the regulation of industry. In addition, the Offshore Petroleum Boards are crucial to the

---

203. The Memoranda of Understanding that have been provided for in the Canadian legislation should help to reduce interdepartmental rivalry and promote a one-window approach to the regulation of industry.

operation of the whole system. In the case of the Canada - Nova Scotia Offshore Petroleum Board, it is likely that some of its members will come from the affected government departments, but in the case of the Canada-Newfoundland Offshore Petroleum Board, relationships will have to be fostered directly between the Board and the governments. In Canada, therefore, it seems that even more time will have to be devoted to consultation and discussion than in Australia, because of the involvement of an extra institutional layer.

In Canada, COGLA is in an obvious position to take the lead in lubricating relations between the various groups. At the moment there are not any formal mechanisms to ensure ongoing communications between the interested parties, but accomplishments in this regard at an informal level have been noteworthy. This approach may suffice as long as the individuals involved are effective communicators, but some thought should be given to the desirability of developing a more formal mechanism for ensuring ongoing consultation and communication. This would seem to be particularly helpful in ensuring that the Boards themselves stay in close communication, which will become increasingly necessary with the establishment of new arrangements with British Columbia and the two northern territories.

## The Clarity of Legislative Authority

The Australian solution to problems of offshore jurisdiction seems to have left few, if any, problems of determining which level of government has legislative authority over particular matters. The Commonwealth PSLA governs petroleum operations beyond the three-mile state coastal waters; otherwise, legislative jurisdiction in the offshore largely follows the Constitution's general division of authority. A degree of overall Commonwealth control over the adjacent areas is retained through s.9(8) of the PSLA, which enables state law to be made inapplicable by virtue of regulations passed pursuant to the PSLA.

The situation in Canada is less clear-cut, for several reasons.

First, in relation to the east coast, the jurisprudence has only resolved finally the matter of continental shelf rights off Newfoundland. For other areas, mirror legislation may be the best solution, but if the mirror statutes are not in perfect harmony at a given moment, it will be unclear from a legal point of view which statute governs. Second, the absence of an offshore laws application statute leaves considerable uncertainty about what legal principles would apply to certain circumstances. Third, there remains some uncertainty about which labour laws apply to personnel on supply and support craft. Fourth, although the enabling legislation has purported to make it clear that provincial social laws apply offshore, provisions in the enabling legislation that permit certain regulations to override those laws have given rise to confusion. Finally, general legal problems concerning the regulation of MODUS do not seem to have been entirely resolved as yet under the Canadian system.

### Fiscal Matters

Intergovernmental fiscal relations have been and remain problematic under the Australian system. Undoubtedly, this is partly a reflection of the Australian states' dissatisfaction with their overall fiscal situation under the Constitution. Even so, it is possible to identify a number of reasons why the fiscal arrangements for offshore petroleum have given rise to extreme difficulties. One is the nature of the royalty mechanism which has caused innumerable technical problems in application. The Commonwealth's appropriation of much of Victoria's potential offshore revenue through the crude oil levy has been a major irritant. Royalty-sharing between the two levels of government has created many administrative and legal complications and necessitated amendments to the PSLA. The Commonwealth's decision to apply cash bidding and the RRT to the Commonwealth adjacent areas despite state opposition and, even more importantly, in the absence of agreement as to how to share these revenues in the face of the *1979 OCS*, are a further example of the continuing fiscal problems that have plagued Australia's regime. Some of these issues have left industry trapped in the middle of intergovernmental disputes. In some situations the disagreements have required industry's participation in litigation and government inquiries.

While the Canadian system is much younger, so far it appears that most of these kinds of problems have been avoided. For instance, provincial control over the royalty system's share and proceeds removes the need for duplicative calculation and verification procedures.[204] The fact that one regime and not two applies to the entire area offshore of a particular province removes the uncertainties that industry has experienced in Australia about which fiscal regime will apply to fields that overlap federal-state boundaries. However, as will be seen in the section entitled "Boundaries", this remains a potential concern in Canada as between adjoining provinces and, perhaps more importantly, in relation to arrangements for the North.

Despite the relative cleanness of the royalty arrangements so far in Canada, in a federal state there will always be a need for coordination between the two levels of government about the overall tax and royalty regime's impact on industry, especially in the context of high-cost, high-risk projects. The negotiations over the Hibernia project demonstrate that in Canada, to date, the affected governments have managed to sort out their arrangements ahead of time, finessing some of the problems that persist in Australia. It remains to be seen whether the royalty systems

---

204. It appears that federal-territorial revenue-sharing is contemplated, however, in relation to Arctic offshore areas. The *Northwest Territories Northern Accord Agreement in Principle*, *supra*, note 2, makes provision for two forms of revenue-sharing. First, the two territories themselves are required to negotiate an agreement relating to the sharing of oil and gas resource revenue from the Beaufort Sea. Second, it is provided that, if territorial oil and gas revenues achieve a particularly high level, to be determined, an increasing portion of incremental revenues will be retained by the Government of Canada. These arrangements imply a need to design revenue-sharing mechanisms that have not been required under the east coast Accords.

developed by Nova Scotia and Newfoundland will give rise to the industry-government disputes concerning calculations that have been endemic in Australia.

Boundaries

The question of boundaries is more significant under Australia's system than under Canada's, for two reasons. First, all the states have coastlines, so boundaries had to be drawn between each jurisdiction. Second, the *1979 OCS* necessitated the drawing of a second set of boundaries, namely the boundaries between state coastal waters and the Commonwealth adjacent areas. Settling both types of boundaries proved to be time-consuming and politically sensitive. The existence of so many boundaries has given rise to a host of practical, legal and theoretical issues, particularly in relation to prospects or fields that transverse one or more set of boundaries. As the notion of a "Common Mining Code" becomes less of a reality than it was when the regime was designed in 1967, such problems will be exacerbated. The result is a myriad of problems and uncertainties for industry including the question of what rights issuance and fiscal regimes will apply to transboundary areas. Frustrations and costs of coping with more than one registry system have also resulted.

So far, this issue has been addressed in Canada only through the legislative procedures that permit negotiation or arbitration of interprovincial offshore boundaries. Despite the existence of some permits in areas offshore the east coast that could fall between provinces and the Northwest Territories, the absence of promising finds in these areas has made boundary delineation a matter of way low priority. If and when the boundaries are set in Canada, however, an array of other issues will need to be addressed, including the allocation of petroleum to overlapping pools and fields and the question of what laws, management regimes, and fiscal arrangements will apply in such circumstances. It seems that the necessity to address these issues may arise first in the context of the northern offshore areas.[205] In any event, these questions should be faced in advance of real problems in order to avoid the sort of *ad hoc* negotiations that seem to have set the pattern in Australia. It will clearly be important for the federal government to play a strong leadership role in regard to such matters.

---

205. As indicated *id.*, the Northwest Territories and the Yukon governments must together negotiate arrangements for the Beaufort Sea.

## 5. CONCLUSIONS

For over two decades offshore petroleum management arrangements have strained the fabric of two federal states, Canada and Australia. Despite some resort to the courts, the solutions in both countries have resulted from a process of intergovernmental negotiation. However, as this study has shown, problems remain and there is room for improvement in each. The experience in these two federal states over the past two decades suggests the following conclusions.

At a general level, it is suggested that the form taken by an intergovernmental agreement can have important implications, especially for the party or parties in a weaker bargaining position. Thus, the general and, to some extent, vague language of the *1979 OCS* has worked to the detriment of the Australian states in relation to the changing distribution of authority as between the JA and the DA and to evolving fiscal relations. In contrast, the Canadian Accords, being relatively detailed, will make it difficult for either level of government to promote changes to the system based upon the argument that the Accords are unclear or did not contemplate certain issues. That the vague nature of parts of the *1979 OCS* seem to have worked against the interests of the Australian states is an important lesson to be remembered by British Columbia and the northern territories as their offshore negotiations move forward.

A related point is that it seems desirable to address, through such intergovernmental agreements, as many issues as can be initially identified: experience demonstrates that difficult issues do not disappear simply because they are ignored or avoided during the course of such negotiations. For example, although the *1979 OCS* said very little about fiscal arrangements, Commonwealth

-state revenue-sharing problems continue to arise and must be confronted. A clearer framework for these arrangements in the first instance might have made the process of dealing with them less acrimonious.

In both Canada and Australia, preliminary discussions are underway concerning the design of a cooperative regime for offshore mining. While some of the matters to be addressed in these discussions are clearly different than those that relate to petroleum, officials should nonetheless be aware that certain problems created by the offshore petroleum regimes can be avoided as an offshore mining framework is developed.

In Canada's east coast Accords, it appears that some of the problems endemic to the Australian system have been avoided (such as revenue-sharing arrangements) or are not of immediate importance (such as cross-boundary arrangements). These matters may prove to be extremely significant in a northern context, however, in particular in relation to territorial management and revenue arrangements for the Beaufort Sea and the prospect of federal-territorial revenue-sharing mechanisms. In regard to these issues, both federal and territorial negotiators would be wise to note some of the issues currently arising under the Australian system and to address these problems in advance.

In addition to these matters, there are a number of important legal and administrative problems that remain to be dealt with under Canada's offshore petroleum regime. Consideration should immediately be given to an overall offshore laws application statute. If constitutional entrenchment of the Accords is pursued, care will need to be taken to do so in a manner that does not foreclose the ability to alter the petroleum legislative framework as dictated by experience. The decision-making mechanisms need to be streamlined and rationalized. In particular, the role of the Oil and Gas Committee must be reconsidered and, if experience suggests it is necessary, other decision-making arrangements may have to be simplified and harmonized. At the least, in designing the arrangements for western and northern offshore areas, it should be possible to organize decision-making in ways that are less complex than those found in the east coast Accords.

Perhaps most importantly, and particularly as arrangements are finalized with the British Columbia and territorial governments, attention must be paid to the coordination of Canada's entire offshore petroleum regime. The retention of a "common core" of legislation for offshore petroleum without resulting in an overly rigid, unamendable system will be a major challenge for all the governments. The overall responsibility for this must lie with the federal government. Having made the political decision to share authority with the coastal provinces and the territories, the federal government must now ensure the workability of the complex, cumbersome system it has helped put in place. As COGLA's role as land manager diminishes, it is the logical agency to take the lead in this regard. The 1990s and beyond will test its capacity to respond to this significant challenge.

In contrast to Canada, Australia's offshore petroleum regime is relatively mature and has proven relatively successful. As a result, there are limited suggestions to be made for improving it. In the fiscal area, however, Canada's success so far at designing a system that appears to be workable may offer a model as it suggests that, even in a federal state, a genuine commitment to cooperation can help resolve difficult problems in advance.

As for industry, it is obvious that the conduct of business in a federal state carries with it certain built-in costs, such as the need to be sensitive to intergovernmental politics and to ensure the promotion of industry interests in the face of intergovernmental disputes. It may also be that the conduct of business in a federal state is accompanied by certain benefits, such as the opportunity to enlist the support of one level of government in disputes with the other.

# TABLE A - PART I

## Atlantic Accord Act

### CNOPB Decisions that are not Fundamental or Subject to Interaction with the Oil and Gas Committee, Approval, Consultation or Ministerial Directives

| Section | Nature of Decision |
|---|---|
| 51 | - delegate its powers or performance of its duties |
| 52 | - appoint and fix terms of reference of advisory bodies |
| 53(2) | - designate interest holders as representatives of interest owner |
| 56(1)(b) | - prohibit commencement or continuation of any work or activity in the case of dangerous or extreme weather conditions affecting health or safety of people or equipment |
| 57(2) | - issue interests that are mandatory under Part II of the Act |
| 59(2) | - select a bid in response to a call for bids |
| 77(1) | - provide information or documentation relating to a significant discovery |
| 80(2) | - authorize petroleum production for use in exploration, drilling or development of petroleum |
| 86(1) | - issue licence for subsurface storage of petroleum or any other substance |
| 121(1) | - order that certain classes of disputes between holders of interests be submitted to arbitration |
| 126(1) | - prescribe forms |
| 138(1) | - issue operating licence or, in certain circumstances, authorize proposed work or activity |
| 138(3) | - suspend or revoke an operating licence for non-compliance, contravention or default |

## Atlantic Accord Act (continued)

| Section | Nature of Decision |
|---|---|
| 139(4)(b), (5) | - approve Part II (Technical Information) of development plan or alteration or amendment thereto |
| 140 | - designate Chief Conservation Officer |
| 141(1) | - establish Oil and Gas Committee |
| 147 | - refer questions or matters to Oil and Gas Committee for a report or recommendation |
| 151.1 | - issue and publish guidelines and interpretation notes with respect to development plan approval process (s.139) and certain regulations |
| 154(1) | - consent to commence prosecution for offence of committing waste |
| 163(1) | - require a person authorized to carry out work or activity to provide proof of financial responsibility |
| 163(2), (3) | - determine manner, conditions, procedures and amount of payment of claims, in the absence of regulations |
| 164(3) | - promote and monitor compensation policies for fishermen suffering damages of a non-attributable nature |
| 165(1) | - direct inquiry into death, injury or danger to public safety or environment resulting from offshore production activity |
| 167(2) | - enter into a pooling agreement |
| 172(2) | - enter into a unit agreement |
| 186 | - review orders made by Oil and Gas Committee under Part III of the Act (Production and Conservation) and vary or rescind such decision |
| 188 | - appoint conservation engineers |
| 202(1) | - commence and maintain an action to enjoin contraventions of Part III of the Act (Production and Conservation), the regulations, or orders |

# TABLE A - PART II

## Atlantic Accord Act

### CNOPB Fundamental Decisions*

| Section | Nature of Decision |
|---|---|
| 54(1) | - prohibit issuance of interests in respect of certain offshore areas |
| 56(1)(a) | - prohibit interest owner from commencing or continuing any work or activity in the case of an environmental or social problem of a serious nature |
| 57(1) | - issue interests unless issuance is mandatory |
| 58(2) | - make a call for bids in relation to the issue of an interest in Crown reserve areas |
| 61(1) | - issue an interest in relation to any Crown reserve area without making a call for bids, in particular circumstances |
| 67(1) | - agree with interest owner on terms and conditions to be contained in an exploration licence |
| 68(1) | - agree with interest owner on amendment of provisions of an exploration licence |
| 68(3) | - consolidate two or more exploration licences into a single exploration licence |
| 73(2), (3) | - issue a significant discovery licence respecting Crown reserve area after making a call for bids |
| 73(4) | - agree with interest owner on terms and conditions to be contained in a significant discovery licence |
| 76(1) | - issue a drilling order to interest owner of a significant discovery area (subject to a hearing by the Oil and Gas Committee if one was requested: see Table A, Part IV) |

*The procedure for fundamental decision-making is set out in ss.31-40 of the *Atlantic Accord Act*. All fundamental decisions are subject to the power of the federal and provincial ministers to issue joint written directives (Section 42(1)(a). See Table A, Part III)

## Atlantic Accord Act (continued)

| Section | Nature of Decision |
|---|---|
| 79(1) | - give notice or order to reduce term of interest if production is not commenced in a commercial discovery area |
| 79(3) | - reduce the term of an interest in a commercial discovery area (subject to a hearing by the Oil and Gas Committee if one was requested: see Table A, Part IV) |
| 79(6) | - extend the time specified in an interest reduction order or revoke the order |
| 81(1)(b) | - issue a production licence to one interest owner in respect of more than one area or to more than one interest owner in respect of one or more areas |
| 81(2), (3) | - issue a production licence in respect of a commercial discovery area extending to Crown reserve after a call for bids |
| 81(4) | - agree with interest owner on terms and conditions to be contained in a production licence |
| 82 | - consolidate two or more production licences |
| 84(4) | - extend the term of a production licence in certain cases |
| 123(2) | - cancel interest or portion of interest for failure to comply with notice to comply (subject to a hearing by Oil and Gas Committee if one was requested: see Table A, Part IV) |
| 130(1) | - negotiate an exploration licence with interest owner of a former permit, special renewal permit or former exploration agreement |
| 131(1) | - negotiate an exploration licence with interest owner of a former lease |
| 133(2) | - agree with interest owners on terms and conditions of a consolidated exploration licence |
| 139(4)(a), (5) | - approve Part I or amendment or alteration of Part I of a development plan (general approach of developing a pool or field) |

# TABLE A - PART III

## Atlantic Accord Act

### CNOPB Decisions Subject to Joint Ministerial Directives (s.42)

| Section | Nature of Decision |
|---|---|
| 42(1)(a) | - all Fundamental Decisions (see Table A, Part II) |
| 42(1)(c), 44 | - conduct a public review in relation to any potential development of a field or pool |
| 42(1)(d), 45(6) | - approval of Canada-Newfoundland benefits plan |
| 42(1)(b), 56(1)(b) | - decisions to prohibit commencement or continuation of work or activity in the case of dangerous or extreme weather conditions affecting health or safety of people or safety of equipment |
| 42(1)(e) | - studies to be conducted by the Board |

# TABLE A - PART IV

## Atlantic Accord Act

### CNOPB Decisions Subject to Interaction With the Oil and Gas Committee*

| Section | Nature of Decision |
|---|---|
| 71(1), (2) | - make a declaration of significant discovery on application of interest holder or on Board's initative |
| 71(4) | - amend or revoke a declaration of significant discovery following further drilling |
| 76(1) | - issue a drilling order to interest owner of significant discovery area (also an FD: see Table A, Part II) |
| 78(1), (2) | - make a declaration of commercial discovery on application by interest holder or on Board's initiative |
| 79(3) | - make an order reducing the term of an interest in commercial discovery area (also an FD: see Table A, Part II) |
| 123(2) | - cancel an interest or portion of an interest for failure to comply with notice to comply (also an FD: see Table A, Part II) |

*The Oil and Gas Committee's role is set out in s.124 of the Act.

# TABLE A - PART V

## Atlantic Accord Act

### Matters Requiring the CNOPB to Consult With or Obtain Approval of One or Both Ministers

| Section | Nature of Decision |
|---|---|
| 43(1) | - prepare plan regarding calls for bids and issuance of interests (approval of minister with fundamental decision-making power required) |
| 45(5) | - review of Canada-Newfoundland benefits plan (consultation with ministers required) |
| 46 | - conclude memoranda of understanding with federal and provincial departments (participation of ministers required) |
| 101(3) | - appoint a member to the Environmental Studies Management Board (on recommendation of provincial minister) |
| 139(1) | - authorize work or activity relating to developing a field or pool unless the Board has approved a development plan (approval of ministers required) |
| 142(3) | - provide staff to Oil and Gas Committee from other than staff of Board (approval of minister required) |

# TABLE A - PART VI

## Atlantic Accord Act

### Matters Where the CNOPB Must act on the Request of the Federal Minister or not act Unless Federal Minister is Satisfied

| Section | Nature of Decision |
|---|---|
| 40(1) | - cause production of suitable crude oil and equivalent substances to be increased, consistent with good oilfield practice, in the event of a supply shortfall |
| 40(2) | - take such measures as are necessary to comply with Canada's obligations pursuant to the *1974 Agreement On An International Energy Program* and as are fair and equitable in relation to other hydrocarbon producing regions of Canada |
| 89, 90, 91, 95, 96 | - Canadian ownership requirements must be satisfied prior to issue of production licence (production licence terms and conditions are also a fundamental decision. See Table A, Part II) |

# TABLE B

## Atlantic Accord Act

### Federal Regulations re Newfoundland to be Made Only With Approval of Provincial Minister

| Section | Nature of Decision |
|---|---|
| 5(1) | - Regulations prescribing lines enclosing areas adjacent to the Province for the purpose of the definition of "offshore area" in s.2(a) of Act |
| 41(7) | - Regulations for carrying out the purposes and provisions of s.41 (Regional Security of Supply) |
| 67(2) | - Regulations prescribing terms and conditions required to be included in exploration licences |
| 118 | - Regulations for carrying out the purposes and provisions of Division VIII (Transfers, Assignments and Registration) |
| 122(1) | - Regulations for carrying out the purposes and provisions of ss.120 and 121 (Arbitration) |
| 125(1) | - Regulations for carrying out the purposes and provisions of Part II (Petroleum Resources) |
| 149(1) | - Regulations respecting the exploration and drilling for and the production and conservation, processing and transportation of petroleum |
| 152(5) | - Regulations prescribing a work or a class of works for the purpose of the definition of "marine installation or structure" and prescribing any provincial Act for the purpose of s.152(2) or excluding any such Act from the application of that subsection |
| 160(4) | - Regulations authorizing the discharge, emission or escape of petroleum |
| 203 | - Regulations deemed necessary for carrying out the purposes of Part III (Production and Conservation) |

# TABLE C - PART I

## Nova Scotia Accord Act

### CNSOPB Decisions that are not Fundamental or Subject to Interaction with the Oil and Gas Committee, Approval, Consultation or Ministerial Directives

| Section | Nature of Decision |
|---|---|
| 54 | - delegate its powers or performance of its duties |
| 55 | - appoint and fix terms of reference of advisory bodies |
| 56(2) | - designate interest holders as representative of interest owner |
| 59(1)(b) | - prohibit commencement or continuation of any work or activity in the case of dangerous or extreme weather conditions affecting health or safety of people or equipment |
| 60(2) | - issue interests that are mandatory under ss.76(1) or 84(1)(a) |
| 62(2) | - select a bid in response to a call for bids |
| 80(1) | - provide information or documentation relating to a significant discovery |
| 83(2) | - authorize petroleum production for use in exploration, drilling or development of petroleum |
| 124(1) | - order that certain classes of disputes between holders of interests be submitted to arbitration |
| 129(1) | - prescribe forms |
| 142(1) | - issue operating licence or in certain circumstances authorize proposed work or activity |
| 142(3) | - suspend or revoke operating licence for non-compliance, contravention or default |
| 143(4)(b), (5) | - approve Part II (Technical Information) of development plan or alteration or amendment thereto |
| 144 | - designate Chief Conservation Officer |

## Nova Scotia Accord Act (continued)

| Section | Nature of Decision |
|---|---|
| 145(1) | - establish Oil and Gas Committee |
| 151 | - refer questions or matters to Oil and Gas Committee for a report or recommendation |
| 156 | - issue and publish guidelines and interpretation notes with respect to certain matters |
| 159(1) | - consent to prosecution for offence of committing waste |
| 168(1) | - require a person authorized to carry out work or activity to provide proof of financial responsibility |
| 168(2) | - determine manner, conditions, procedures and amount of payment of claims, in the absence of regulations |
| 169(3) | - promote and monitor compensation policies for fishermen suffering damages of a non-attributable nature |
| 170(1) | - direct inquiry into death, injury or danger to public safety or environment resulting from offshore production activity |
| 172(2) | - enter into a pooling agreement |
| 177(2) | - enter into a unit agreement |
| 191 | - review orders made by Oil and Gas Committee under Part III of the Act (Production and Conservation) and vary or rescind such decision |
| 193 | - appoint conservation engineers |
| 207(1) | - commence and maintain an action to enjoin contravention of Part III of the Act (Production and Conservation), the regulation or orders |

# TABLE C - PART II

## Nova Scotia Accord Act

### CNSOPB Fundamental Decisions*

| Section | Nature of Decision |
|---|---|
| 59(1)(a) | - prohibit interest owner from commencing or continuing any work or activity in the case of an environmental or social problem of a serious nature |
| 60(2) | - issue interests unless issuance is mandatory |
| 61(2) | - make a call for bids in relation to the issue of an interest in Crown reserve areas |
| 64(1) | - issue an interest in relation to any Crown reserve area without making a call for bids, in particular circumstances |
| 70(1) | - agree with interest owner on terms and conditions to be contained in an exploration licence |
| 71(1) | - agree with interest owner on amendment of provisions of an exploration licence |
| 71(3) | - consolidate two or more exploration licences into a single exploration licence |
| 76(2), (3) | - issue a significant discovery licence respecting Crown reserve area after making a call for bids |
| 76(4) | - agree with interest owner on terms and conditions to be contained in a significant discovery licence |
| 79(1) | - issue a drilling order to interest owner of a significant discovery area (subject to a hearing by the Oil and Gas Committee if one was requested: see Table C, Part IV) |
| 82(1) | - give notice to reduce term of interest if production is not commenced in a commercial discovery area |

*The procedure for fundamental decisions is set out in ss.32-37 of the *Nova Scotia Accord Act*. All fundamental decisions are subject to the power of the federal and provincial minister to issue joint written directives (s.41(a). See Table C, Part III).

## Nova Scotia Accord Act (continued)

| Section | Nature of Decision |
|---|---|
| 82(3) | - reduce the term of an interest in a commercial discovery area (subject to a hearing by the Oil and Gas Committee if one was requested: see Table C, Part IV) |
| 82(6) | - extend the time specified in an interest reduction order or revoke the order |
| 84(1)(b) | - issue a production licence to one interest owner in respect of more than one area or to more than one interest owner in respect of one or more areas |
| 84(2), (3) | - issue a production licence in respect of a commercial discovery area extending to Crown area after a call for bids |
| 84(4) | - agree with interest owner on terms and conditions to be contained in a production licence |
| 85 | - consolidate two or more production licences |
| 87(4) | - extend the term of a production licence in certain cases |
| 89(1) | - issue a licence for subsurface storage of petroleum or other substances |
| 126(2) | - cancel interest or portion of interest for failure to comply with notice to comply (subject to a hearing by Oil and Gas Committee if one was requested: see Table C, Part IV) |
| 133(1) | - negotiate an exploration licence with interest owner of a former permit, special renewal permit or former exploration agreement |
| 134(1) | - negotiate an exploration licence with interest owner of a former lease |
| 136(2) | - agree with interest owners on terms and conditions of a consolidated exploration licence |
| 143(4)(a), (5) | - approve Part I or amendment or alteration of Part I of a development plan (general approach of developing a pool or field) |

# TABLE C - PART III

## Nova Scotia Accord Act

### CNSOPB Decisions Subject to Ministerial Directives (s.41)

| Section | Nature of Decision |
| --- | --- |
| 41(a) | - all Fundamental Decisions (joint) (See Table C, Part II) |
| 41(b), 45(6) | - approval of Canada - Nova Scotia benefits plans (joint) |
| 41(c),44 | - conduct public reviews in relation to the exercise of its powers or performance of its duties (joint) |
| 41(d) | - studies to be conducted by the Board (joint) |
| 41(e) | - advice with respect to policy issues to be given by the Board to the ministers (joint) |
| 41(2) | - call for bids in relation to particular portions of the offshore (either) |
| 41(3) | - Canadian ownership rate relating to any interest that may be issued (federal) |
| 41(4) | - Fundamental Decisions relating to the Bay of Fundy or Sable Island (provincial) |

# TABLE C - PART IV

## Nova Scotia Accord Act

### CNSOPB Decisions Subject to Interaction With the Oil and Gas Committee*

| Section | Nature of Decision |
|---|---|
| 74(1), (2) | - make a declaration of significant discovery on application of interest holder or on Board's initiative |
| 74(4) | - amend or revoke a declaration of significant discovery following further drilling |
| 79(1) | - issue a drilling order to interest owner of significant discovery area (also a Fundamental Decision: see Table C, Part II) |
| 81(1), (2) | - make a declaration of commercial discovery on application by interest holder or on Board's initiative |
| 82(3) | - make an order reducing the term of an interest in a commercial discovery area (also a Fundamental Decision: see Table C, Part II) |
| 126(2) | - cancel an interest or portion of an interest for failure to comply with notice to comply (also a Fundamental Decision: see Table C, Part II) |

*The Oil and Gas Committee's role is set out in s.127 of the *Nova Scotia Accord Act.*

# TABLE C - PART V

## Nova Scotia Accord Act

### Matters Requiring the CNSOPB to Consult With or Obtain Approval of One or Both Ministers

| Section | Nature of Decision |
|---|---|
| 45(5) | - review of Canada - Nova Scotia benefits plan (consultation with ministers required) |
| 46 | - conclude memoranda of understanding with federal and provincial departments and agencies (participation of ministers required) |
| 103(3) | - appoint a member to the Environmental Studies Management Board (on recommendation of provincial minister) |
| 143(1) | - authorize work or activity relating to developing a field or pool unless the Board has approved a development plan (approval of ministers required) |
| 146(3) | - provide staff to Oil and Gas Committee from other than staff of Board (approval of ministers required) |

# TABLE C - PART VI

## Nova Scotia Accord Act

### Matters Where the CNSOPB Must act on the Request of the Federal Minister or not act Unless Federal Minister is Satisfied

| Section | Nature of Decision |
|---|---|
| s.38(2) | - take such measures as are authorized by the federal minister to meet Canada's obligations pursuant to the *1974 Agreement On An International Energy Program* and as are fair and equitable in relation to other hydrocarbon producing regions of Canada |
| 91, 92, 93, 97, 98 | - Canadian ownership requirements must be satisfied prior to issue of production licence (production licence terms and conditions are also a Fundamental Decision. See Table C, Part II) |

# TABLE D

## Nova Scotia Accord Act

### Federal Regulations re Nova Scotia to be Made Only With Approval of Provincial Minister

| Section | Nature of Decision |
| --- | --- |
| 5(1) | - regulations amending description of limits, as set out in Schedule 1, for purposes of defining "offshore area" |
| 17(4) | - regulations prescribing terms and conditions of indemnification |
| 35(8) | - regulations prescribing time and manner of application to review and decision to be made regarding a veto made under the section |
| 39(7) | - regulations for carrying out the purposes and provisions of s.39 (Regional Security of Supply) |
| 45(7) | - regulations prescribing time and manner of submission of any Canada - Nova Scotia benefits plan and information to be contained therein |
| 67 | - regulations of general application for the purpose of carrying out s.61 (calls for bids, issuance of interests in relation to Crown reserve areas) |
| 70(2) | - regulations prescribing terms and conditions to be included in exploration licences |
| 121 | - regulations for carrying out purposes and provisions of Division VIII (transfers, assignments and registration) |
| 125(1) | - regulations for carrying out the purposes and provisions of ss.123 and 124 (Arbitration and Operating Agreements) |
| 128(1) | - regulations for carrying out the purposes and provisions of Part II (Petroleum Resources) |
| 153(1) | - regulations respecting exploration and drilling for and the production and conservation, processing and transportation of petroleum |

## Nova Scotia Accord Act (continued)

| Section | Nature of Decision |
|---|---|
| 157(5) | - regulations prescribing a work or class of works for the purpose of the definition of "marine installation or structure" in s.157(1) and regulations prescribing for or excluding from s.157(2) any Act of the Legislature of the Province |
| 165(4) | - regulations authorizing the discharge, emission or escape of petroleum of such types, in such quantities, at such locations, under such conditions and by such persons as the regulations specify |
| 208 | - regulations for carrying out the purposes and provisions of Part III (Production and Conservation) and in particular for defining terms "oil" and "gas" |
| 245 | - regulations for carrying out the purposes and provisions of Part VII (Drilling Fund) |
| 248 | - regulations prescribing anything to be prescribed for the purposes of Part VII (Crown Share Adjustment Payments) |

# TABLE E

## Powers of Australia's Joint Authority, Pursuant to PSLA

| Section | Nature of Decision |
|---|---|
| *Part III* | |
| *Division 1* | |
| *Preliminary* | |
| 18(1) | - declare a block to be reserved |
| *Division 2* | |
| *Exploration Permits for Petroleum* | |
| 22(1), (4) | - grant exploration permit |
| 22A(1), (6) | - invite applications by way of cash bidding |
| 22B(1), (2), (3), (5) | - grant permit or reject applications for exploration permit |
| 25(1), (2), (4), (5), 27 | - grant permit applied for under s.23, or reject applications for exploration permit |
| 32(1), (2), (6) | - grant/refuse to grant renewal of an exploration permit |
| 33(1), (2), (2A) | - grant an exploration permit subject to such conditions thought fit, except a permit granted under s.22B which may not be subject to minimum work and expenditure requirements |
| 36(6), (8), (9) | - nomination by permittee or JA of block(s) as a location |
| 37(1)(b), 37(2) | - declare nominated block(s) as a location |
| 37(3), (4), (5), (6) | - revoke or vary a declaration of alocation |
| *Division 2A* | |
| *Retention Leases for Petroleum* | |
| 38B(1), (2), (5) | - grant/refuse to grant a retention lease |
| 38E(1), (2) | - cancel lease |

## Powers of Australia's Joint Authority, Pursuant to PSLA (continued)

| Section | Nature of Decision |
|---|---|
| 38G(1), (2), 38H | - grant/refuse to grant renewal of a lease |
| *Division 3* *Production Licences* | |
| 42(1), (2) | - determine royalty rate after giving applicant an opportunity to confer with D.A |
| 43(1), 44(2) | - grant production licence in respect of application made under ss.39A, 40, 40A or 40B |
| 49(1), (2), (4), 50 | - reject application or grant licence in respect of a surrendered, cancelled or determined interest |
| 51(4) | - grant production licence for individual block(s) |
| 55(1), (2), (3), (5), (6), (9)(b) | - grant/refuse to grant licence renewal |
| 58 | - give licencee directions regarding recovery and production of petroleum |
| 59(3), (4), (5), (6), (7), (8) | - give licencee directions regarding unit development to secure more effective recovery of petroleum |
| 59(12) | - obtain approval of appropriate State or Territorial authority before giving approval or issuing directions in respect of a pool located in more than one jurisdiction |
| *Division 4* *Pipeline Licences* | |
| 64(3) | - reject application for pipeline licence by person other than licencee |
| 65(1) to (5), (10) | - grant/refuse to grant pipeline licence |
| 65(8) | - specify route for pipeline under consideration |
| 67(1) | - specify term of pipeline licence |

## Powers of Australia's Joint Authority, Pursuant to PSLA (continued)

| Section | Nature of Decision |
|---|---|
| 69(1), (2), (6) | - grant/refuse to grant renewal of pipeline licence |
| 70(1) | - grant pipeline licence subject to any conditions thought fit |
| 72 | - direct that changes be made in a pipeline or pipeline facilities and vary licence accordingly |
| 73 | - direct pipeline licencee to be a common carrier of petroleum |
| *Division 5*<br>*Registration of Instruments* | |
| 78(1) | - approve a transfer of title, subject to s.78(4), (6) |
| 81(10) | - approve or refuse to approve a dealing which would create or assign interests in existing rights and titles, except a transfer under s.78 and subject to ss.81(5), (6) |
| *Division 6*<br>*General* | |
| 103(1) | - vary, suspend or exempt interest holder from compliance with any conditions to which interest was made subject |
| 103(3) | - extend the term of a permit or lease |
| 103A(1), (3) | - suspend/terminate suspension of all or any rights conferred by a permit or lease and extend the term of permit or lease |
| 105 | - cancel permit, lease, licence or pipeline licence in the appropriate circumstances |

# TABLE F

## Powers of Australia's Designated Authority, Pursuant to PSLA

| Section | Nature of Decision |
|---|---|
| *Part III* <br> *Division 1* <br> *Mining for Petroleum* | |
| 15(1) | - designate any powers to a person |
| *Division 2* <br> *Exploration Permits for Petroleum* | |
| 20(1) | - invite applications, on own initiative or at request of Commonwealth Minister, for grant of a permit in respect of a block or blocks |
| 20(2) | - direct that restrictions on block numbers and location provided by ss.21(2), (3) do not apply |
| 21(4) | - require applicant to furnish further information |
| 22(3) | - extend period for request of permit grant or lodging security |
| 23(1), (5) | - invite applications for grant by JA of a permit in respect of a block or blocks where the interest has terminated and require applicant to furnish further information |
| 24(3) | - refund deposit |
| 26(1) | - extend period for request of permit grant or lodging security |
| 30(3) | - receive an application for renewal of a permit within 3 months of expiration of the permit |
| 31(5) | - determine number of blocks and give directions concerning the blocks for which an application for renewal of a permit may be made |

## Powers of Australia's Designated Authority, Pursuant to PSLA (continued)

| Section | Nature of Decision |
|---|---|
| 31(6) | - direct that restrictions on blocks provided by ss.31(3), (4) do not apply |
| 34(2) | - direct permit holder of petroleum discovery area to furnish technical information relating to the discovery |
| 35(1) | - direct permittee to do such things thought necessary to determine the characteristics of the petroleum and petroleum pool |
| *Division 2A*<br>*Retention Leases for Petroleum* | |
| 38A(3), 38F(4) | - require lease applicant to furnish further information |
| 38A(4)(b), 38B(4)<br>38F(3) | - alter application and request deadlines |
| 38J(2) | - direct lessee of petroleum discovery area to furnish technical information relating to the discovery |
| 38K(1) | - direct lessee to do such things thought necessary to determine the characteristics of the petroleum and petroleum pool |
| *Division 3*<br>*Production Licences for Petroleum* | |
| 39A(5), 40(4) | - allow extension of period for licence application |
| 41(2) | - require licence applicant to furnish further information |
| 45(1) | - vary a production licence |
| 46(3), (5), (6) | - revoke declaration of a location |
| 47(1) | - invite applications for grant by JA of a production licence in respect of a block where the interest has terminated |
| 47(7) | - require applicant to furnish further information |
| 48(3) | - refund deposit |

## Powers of Australia's Designated Authority, Pursuant to PSLA (continued)

| Section | Nature of Decision |
|---|---|
| 49(6), 54(3) | - alter application and request deadlines |
| 51(3) | - require applicant to lodge security |
| 57(4) | - exempt licencee from compliance with work required to be carried out |
| 59(11) | - consult with any other DA concerned regarding exploitation of the pool where pool extends beyond state boundaries |
| *Division 4*<br>*Pipeline Licences* | |
| 60 | - consent or refuse consent to construction, re-construction, alteration etc. of pipelines and facilities |
| 61(a)(ii) | - give directions in case of an emergency |
| 62(1), (3) | - direct appropriate person to make alterations to, or to move a pipeline or facilities, or do those things |
| 63 | - declare a terminal station |
| 64(2), 65(9), | - alter application for licence or renewal and request 68(3) deadlines |
| 64(4), 71(3) | - require applicant for licence or renewal to furnish further information |
| 71(4) | - specify time period in which application to be completed |
| 71(5) | - vary pipeline licence |
| 74(1) | - consent, subject to any conditions thought necessary, to cessation of operation of a pipeline |
| *Division 5*<br>*Registration of Instruments* | |
| 76 - 91 | - maintain and administer a register of titles and prospecting authorities |

## Powers of Australia's Designated Authority, Pursuant to PSLA (continued)

| Section | Nature of Decision |
|---|---|
| *Division 6 General* | |
| 94 | - publish notice of grants, expiry, surrender, cancellation etc. of interests |
| 96(2) | - alter 6 month window for commencement of required works |
| 97(2)(d) | - direct that water sources discovered be kept separate |
| 97A | - determine level of insurance to be carried by interest holder |
| 100(1), (2) | - consent to drilling of well near boundary of interest; where interest holder not in compliance, direct that the well be plugged, closed or operated according to the DAs directions |
| 101(1), (3) | - direct registered interest holder with respect to any matter to which regulations may be made, subject to approval of JA if of a standing or permanent nature |
| 101(2C) | - require display of instrument issued under s.101(1) |
| 102(1), (1A) | - do all things required under s.101, on approval of JA if direction was issued by him |
| 103(1A) | - vary or suspend or exempt holder of access or special prospecting authority from compliance with conditions to which authority was subject, subject to any conditions thought necessary |
| 104 | - consent to surrender of an instrument |
| 107(1), (2) | - direct removal or property, plugging or closing off of wells and require provision to be made for conservation, protection and if necessary, restoration of natural resources |
| 108 | - do all or any of things required under s.107(1) |
| 109(1) | - enter agreement for payment by installments of amount in respect of grant of interest |

## Powers of Australia's Designated Authority, Pursuant to PSLA (continued)

| Section | Nature of Decision |
|---|---|
| 110(2) | - remit whole or part of amount payable by reason of late payment of installment |
| 111(3) | - grant or refuse to grant special prospecting authority subject to any conditions thought necessary |
| 111(6B) | - notify special prospecting authority holders of existence and terms of another special prospecting authority held in respect of the same block(s) |
| 111(7)(b), (8) | - cancel special prospecting authority and direct action to be taken on cancellation, surrender or expiry of special prospecting authority |
| 112(3), (7), (8) | - grant, extend, cancel access authority |
| 112(9) | - direct action to be taken on cancellation, surrender or expiry of access authority |
| 113(1), (2) | - remove and dispose of property from relinquished area subject to a direction under s.108 and recover all or part of costs and expenses incurred by DA |
| 114(1)(c) | - allow alternative to cash payment of security |
| 115(1) | - require person believed to be capable of producing certain information to produce such information |
| 116(1) | - administer oath to person examined under s.115 |
| 118 | - make information available to a minister or minister of state and make information public |
| 118(1A) | - make copies of documents received available to Commonwealth Minister on request |
| 119(1) | - declare a safety zone and prohibit entry in zone without consent of the D.A |
| 121(1), (2) | - require interest holder to carry out a survey and to submit a report and any further information necessary |
| 122(1) | - direct that records etc. relating to an interest area or activity be collected and kept, subject to directions issued by the Commonwealth Minister |

## Powers of Australia's Designated Authority, Pursuant to PSLA (continued)

| Section | Nature of Decision |
|---|---|
| 123(1) | - consent to carrying out of scientific investigations in adjacent areas |
| 125 | - appoint inspector for adjacent area |
| 123(1) | - consent to carrying out of scientific investigations in adjacent areas |
| 125 | - appoint inspector for adjacent area |
| *Division 6A*<br>*Safety Zones - Areas to be Avoided* | |
| 140A(4) | - authorize person(s) to perform duties with respect to Divison 6A of Act |
| 140C(2), (3), (4) | - authorize/revoke authorization entry of by vessel into area to be avoided |
| *Division 7*<br>*Transitional Provisions* | |
| 146(10), 147(5) | - enter in Register licence/pipeline licence respecting Barracouta and Marlin fields |
| 148(4) | - require applicant for Barrow Island licence to furnish further information |
| 148(5) | - require applicant for Barrow Island licence to lodge security |
| 148(6) | - grant production licence in respect of Barrow Island interest |
| 149(2), (4) | - amalgamate blocks no longer subject to permit, lease, licence or prescribed instrument with consent of the interest holder concerned |

# INSTITUTE PUBLICATIONS

Following is a complete list of CIRL publications currently available:

*The Offshore Petroleum Regimes of Canada and Australia*, by Constance D. Hunt. 1989. ISBN 0-919269-29-X. $24.00

*The Inuvialuit Final Agreement*, by Janet M. Keeping. 1989. ISBN 0-919269-28-1. $24.00

*Toxic Water Pollution in Canada: Regulatory Principles for Reduction and Elimination with Emphasis on Canadian Federal and Ontario Law*, by Paul Muldoon and Marcia Valiante. 1989. ISBN 0-919269-24-9. 120 p. $22.00

*Interjurisdictional Issues in Canadian Water Management*, by J. Owen Saunders. 1988. ISBN 0-919269-27-3. 130 p. $22.00

*The Framework of Water Rights Legislation in Canada*, by David R. Percy. 1988. ISBN 0-919269-21-4. 103 p. $20.00

*Maritime Boundaries and Resource Development: Options for the Beaufort Sea*, by Donald R. Rothwell. 1988. ISBN 0-919269-24-9. 61 p. $15.00

*A Reference Guide to Mining Legislation in Canada* (Second Edition), by Barry Barton, Barbara Roulston and Nancy Strantz. 1988. ISBN 0-919269-25-7. 123 p. $30.00

*Aboriginal Water Rights in Canada: A Study of Aboriginal Title to Water and Indian Water Rights*, by Richard H. Bartlett. 1988. ISBN 0-919269-23-0. 231 p. $30.00

*Liability for Drilling- and Production-Source Oil Pollution in the Canadian Offshore*, by Christian G. Yoder. Working Paper 12. 1986. ISBN 0-919269-20-6. 84 p. $17.00

*A Guide to Appearing Before the Surface Rights Board of Alberta* (Second Edition), by Barry Barton and Barbara Roulston. Working Paper 11. 1986. ISBN 0-919269-19-2. 124 p. $17.00

*Crown Timber Rights in Alberta*, by N.D. Bankes. Working Paper 10. 1986. ISBN 0-919269-17-6. 128 p. $17.00

*The Canadian Regulation of Offshore Installations*, by Christian G. Yoder. Working Paper 9. 1985. ISBN 0-919269-18-4. 116 p. $17.00

*The Assignment and Registration of Crown Mineral Interests*, by N.D. Bankes. Working Paper 5. 1985. ISBN 0-919269-11-7. 126 p. $17.00

*Oil and Gas Conservation on Canada Lands*, by Owen L. Anderson. Working Paper 7. 1985. ISBN 0-919269-16-8. 122 p. $17.00

*Canadian Maritime Law and the Offshore: A Primer*, by W. Wylie Spicer. Canadian Continental Shelf Law 3; Working Paper 6. 1984. ISBN 0-919269-12-5. 65 p. $13.00

*Public Disposition of Natural Resources*, Essays from the First Banff Conference on Natural Resources Law, April 12-15, 1983; Nigel Bankes and J. Owen Saunders, eds. ISBN 0-919269-14-1. 366 p. (hardcover) $47.00

*Fairness in Environmental and Social Impact Assessment Processes*, Proceedings of a Seminar, The Banff Centre, February 1-3, 1983; Evangeline S. Case, Peter Z.R. Finkle and Alastair R. Lucas, eds. Proceedings 2. ISBN 0-919269-08-7. 125 p. $17.00

*Canadian Electricity Exports: Legal and Regulatory Issues*, by Alastair R. Lucas and J. Owen Saunders. Working Paper 3. 1983. ISBN 0-919269-09-5. 40 p. $9.50

*The International Legal Context of Petroleum Operations in Canadian Arctic Waters*, by Ian Townsend Gault. Canadian Continental Shelf Law 2; Working Paper 4. 1983. ISBN 0-919269-10-9. 76 p. $9.00

*Acid Precipitation in North America: The Case for Transboundary Cooperation*, by Douglas M. Johnston and Peter Finkle. 1983. ISBN 0-919269-05-2. 75 p. $10.00

*Petroleum Operations on the Canadian Continental Margin — The Legal Issues in a Modern Perspective*, by Ian Townsend Gault. Canadian Continental Shelf Law 1; Working Paper 2. 1983. ISBN 0-919269-02-8. $10.00

*Environmental Law in the 1980s: A New Beginning*, Proceedings of a Colloquium, The Banff Centre, November 27-29, 1981; Peter Z.R. Finkle and Alastair R. Lucas, eds. Proceedings 1. ISBN 0-919269-05-2. 233 p. $15.50

*Environmental Regulation — Its Impact on Major Oil and Gas Projects: Oil Sands and Arctic*, by C.D. Hunt and A.R. Lucas. 1980. ISBN 0-919269-00-1. 168 p. $12.95

*Resources: The Newsletter of the Canadian Institute of Resources Law.* ISSN 0714-5918. Quarterly. Free

*1987-88 Annual Report.* Free

## DISCUSSION PAPERS

*Surrounding Circumstances and Custom: Extrinsic Evidence in the Interpretation of Oil and Gas Industry Agreements in Alberta*, by David E. Hardy. 1989. Discussion Paper. $10.00

*Views on Surface Rights in Alberta*, Papers and materials from the Workshop on Surface Rights, Drumheller, 20-21 April 1988; Barry Barton, ed. Discussion Paper. $10.00

*Classifying Non-operating Interests in Oil and Gas*, by Eugene Kuntz; presented at a seminar sponsored by the Faculty of Law and The Canadian Institute of Resources Law, The University of Calgary, 7 April 1988. Discussion Paper. $10.00

**Publications are available from:** Canadian Institute of Resources Law, 430 BioSciences Building, Faculty of Law, The University of Calgary, Calgary, Alberta, Canada T2N 1N4. Telephone (403) 220-3200. Telex 03-821545. Facsimile (403) 282-8325.

**Orders from outside Canada please add $2.00 per book.**

## OTHER PUBLICATIONS

*Canada Energy Law Service.*
ISBN 0-88820-108-7, 5 v.
available from Richard De Boo Publishers
81 Curlew Drive
Don Mills, Ontario, Canada
M3A 3P7

## INSTITUTE PUBLICATIONS AVAILABLE FROM CARSWELL LEGAL PUBLICATIONS

*Trading Canada's Natural Resources*, Essays from the Third Banff Conference on Natural Resources Law, Banff, Alberta, May 6-9, 1987. J. Owen Saunders, ed. 367 p. (hardcover) $75.00*

*Managing Natural Resources in a Federal State*, Essays from the Second Banff Conference on Natural Resources Law, Banff, Alberta, April 17-20, 1985. J. Owen Saunders, ed. 372 p. (hardcover) $70.00*

*Public Disposition of Natural Resources*, Essays from the First Banff Conference on Natural Resources Law, April 12-15, 1983; Nigel Bankes and J. Owen Saunders, eds. ISBN 0-919269-14-1. 366 p. (hardcover) $47.00*

Call Carswell Legal Publications TOLL FREE 1-800-387-5164 (Canada only)
*20% off if you order two or more volumes of Essays from the Banff Conference on Natural Resources Law sponsored by The Canadian Institute of Resources Law.